Leading Digital Transformation - Filtering Through the Maze

Elizabeth T. Atekoja

Abstract

Leading Digital Transformation: Filtering Through the Maze is unique in its approach. It draws from the author's knowledge and experience, combining theories, practical insights, and applications from organizations that have failed and those that have implemented digital transformation successfully.

This book provides practicing leaders and emerging professionals with the theories and best practices to drive digital transformation successfully.

Key Benefits:

- ➢ **Strategic Insight**: Provides seasoned leaders with advanced strategies and frameworks for navigating the complexities of digital transformation, helping them stay ahead in a rapidly evolving landscape.

- ➢ **Practical Guidance:** Offers actionable advice and best practices, equipping leaders and professionals with the tools needed to implement and manage digital initiatives effectively.

- ➢ **Real-World Examples:** Includes practical examples from failed and successful digital transformations, providing valuable lessons and inspiration for both current leaders and future professionals.

- ➢ **Technology Trends and Applications:** Delivers up-to-date information on emerging technologies and their applications, helping leaders make informed decisions and stay current with technological advancements.

- ➢ **Change Management Techniques:** This section covers essential change management techniques to help leaders and professionals drive organizational change smoothly and address resistance effectively.

➢ **Leadership Development:** Focuses on the skills and attributes required for leading digital transformation, offering guidance for developing leadership qualities that are crucial for success in a digital age.

➢ **Cross-Industry Perspectives:** Provides insights from various industries, allowing professionals to understand how digital transformation principles apply across different sectors and adapt strategies accordingly.

➢ **Future-Proofing Skills:** Helps emerging professionals build competencies that are essential for future career success in a digital-first world, ensuring they are prepared for evolving industry demands.

➢ **Enhanced Decision-Making:** Aids both current and future leaders in making strategic decisions by offering a clear understanding of the factors driving digital transformation and their impact on business outcomes.

Preface

Jim Swanson has effectively captured the core of digital transformation in just a few words:

"We discuss automating operations, addressing people, and exploring new business models. Embedded within these areas are data analytics, technologies, and software – all of which serve as enablers rather than drivers. At the heart of it all are leadership and culture. Understanding what digital transformation means for your organization – be it in finance, agriculture, pharmaceuticals, or retail – is crucial." —Jim Swanson, Senior Vice President/CIO and Head of Digital Transformation, Bayer Crop Science.

When I first embarked on exploring the expansive field of digital transformation, I was driven by a profound curiosity and an understanding of its critical significance in today's rapidly changing world. Digital transformation extends far beyond technology; it's about reimagining and redefining the core of how organizations operate, compete, and deliver value. This realization ignited a desire in me to delve deeper into the subject, to grasp its complexities, and to present these insights in a way that would be both accessible and useful to others.

As I conducted research for this book, I drew from my own experiences, as well as those of my colleagues, with the aim of creating a resource that truly supports digital transformation leaders like me. Initially, I contemplated simplifying the content for broader accessibility. However, I quickly recognized that digital transformation is an inherently complex and multifaceted process that demands a deep understanding, keen insights, and a strategic approach.

One of the most rewarding aspects of writing this book has been the continuous learning and growth it has fostered. The process of researching, interviewing industry experts, and analyzing case studies has significantly broadened my knowledge. It has kept me up to date with the latest trends, tools, and strategies, ensuring that the book remains both relevant and timely. This immersion has not only enhanced my professional expertise but has also instilled a mindset of continuous learning and adaptability.

Writing this book has also given me the privilege of connecting with a diverse group of professionals across various industries. Engaging with thought leaders, innovators, and practitioners who are leading the way in digital transformation has been incredibly inspiring. These interactions have provided me with a wealth of perspectives and experiences, deepening my understanding and enabling me to present a more comprehensive and nuanced exploration of the subject.

Transformation is not about taking shortcuts or oversimplifying challenges; it's about embracing the complexities of change, understanding the subtleties of emerging technologies, and developing robust strategies that foster sustainable growth. With this understanding, I set out to write a book that doesn't shy away from the challenges of digital transformation but instead equips leaders with the tools and knowledge they need to navigate these challenges effectively. My goal is to empower leaders to think critically, act decisively, and guide their organizations through transformative changes that will shape the future of their industries.

Moreover, the challenge of distilling complex concepts into clear, actionable insights has been immensely gratifying. It has pushed me to think more critically, to challenge my own assumptions, and to create a more structured and coherent narrative. This intellectual rigor has sharpened my analytical and communication skills, which are invaluable in any professional context.

On a personal level, writing this book has been a journey of self-discovery and empowerment. It has reinforced my belief in the transformative power of technology and innovation. It has also highlighted the importance of resilience, agility, and a forward-thinking mindset in navigating the challenges and opportunities that digital transformation brings.

Ultimately, I hope this book serves as a valuable resource for leaders, entrepreneurs, and professionals who are on their own digital transformation journeys. My aim is to provide practical insights, actionable strategies, and inspiring stories that can guide and support them in driving meaningful and sustainable change within their organizations.

I am thrilled to share this journey with you and hope that it brings you as much value and inspiration as it has brought me. Let's unravel the complexities together, cut through the noise, and pave the way toward sustainable success in the digital age.

Introduction

In an era defined by relentless technological evolution and market disruption, digital transformation has emerged as the cornerstone of organizational survival and growth. Organizations must prioritize agility to remain relevant and sustainable amid continual shifts in the technological landscape. The allure of cutting-edge technologies often tempts organizations to sprint toward implementation without first establishing a solid foundation of strategic alignment.

This rush, fueled by the promise of innovation and competitive edge, can inadvertently lead to disjointed efforts, misaligned priorities, and unrealized potential. In a recent article, KPMG (KPMG LLP, 2024) states that the number of businesses with leadership buy-in for emerging tech has tripled from 10% to 32%.

This book introduces Digital Transformation with a specific emphasis on navigating complexities, discerning between options, and attaining successful outcomes in the swiftly evolving digital era. It delves into the critical nuances of digital transformation, emphasizing the importance of deliberate planning, thoughtful integration, and strategic foresight.

It serves as a guiding compass, steering leaders and practitioners away from hasty technological adoption towards a holistic approach that ensures every innovation aligns with and enhances the broader organizational goals and objectives.

Because of the growing "popularity" of digital transformation, it is imperative to pause and reflect on the essence of digital transformation: it is not merely about adopting the latest technology trends, but rather it encompasses a strategic overhaul where businesses integrate digital technologies into their core operations, aiming not just for efficiency gains but for profound shifts in how value is created and delivered. It also seeks to harmonize where technology, business strategy, and organizational culture seamlessly converge to effectively adapt to and anticipate disruptions.

This introduction sets the stage for our exploration into the multifaceted world of digital transformation. We embark on a journey through the complexities and challenges faced by organizations as they navigate this transformative landscape. From understanding the fundamental principles of digital transformation to strategizing amidst technological mazes, my goal is to provide clarity and guidance.

Throughout this book, we will explore crucial themes for achieving successful digital transformation: defining strategic objectives, overcoming implementation challenges, utilizing filtering strategies to streamline and optimize efforts, leveraging data-driven insights, and cultivating a culture of innovation and agility.

Each chapter is designed to equip you with practical frameworks, real-world examples, and actionable insights that empower you to navigate the complexities of digital transformation effectively.

As we embark on this journey together, let us uncover how organizations can not only adapt to but thrive in the face of technological upheaval. By focusing on alignment, strategic foresight, and a holistic approach to digital adoption, we aim to illuminate the path toward sustainable success in today's rapidly evolving digital economy.

Table1:1 Structure of the book

Introduction	
Part 1- Understanding Digital Transformation	Chapter 1: Unveiling the Core Concepts of Digital Transformation. Chapter 2: Why Digital Transformation Fails.
Part 11 - Strategies and Technologies	Chapter 3: The Maze of choices: Technologies and Strategies Chapter 4: Filtering Strategies for Success
Part 111 – Implementation and Execution	Chapter 5: Leadership and Change Management Chapter 6: Data-Driven Decision Making
Part IV - Advancing Through Collaboration and Future Outlook	Chapter 7: Building Partnerships and Ecosystems. Chapter 8: Measuring Success and Iterating Chapter 9: Future Trends and Considerations
Conclusion	

Examples of Successful Digital Transformations

1. Amazon

Amazon is a prime example of a company that has successfully leveraged digital transformation to become a global leader in e-commerce and cloud computing. Originally an online bookstore, Amazon expanded its digital capabilities to offer a wide range of products and services, including Amazon Web Services (AWS). AWS has revolutionized the way businesses use cloud computing, offering scalable and cost-effective solutions. Amazon's focus on data analytics, personalized customer experiences, and logistics optimization has enabled it to maintain a competitive edge and continually innovate.

2. Netflix

Netflix transformed from a DVD rental service to a leading global streaming platform through digital transformation. Recognizing the shift in consumer preferences towards online streaming, Netflix invested heavily in content delivery networks, data analytics, and personalized recommendation algorithms. This transition not only expanded its user base but also allowed Netflix to create original content based on viewer preferences and trends, making it a dominant player in the entertainment industry.

3. General Electric (GE)

General Electric embarked on a digital transformation journey by integrating industrial equipment with Internet of Things (IoT) technology. GE developed the Predix platform, an industrial IoT software designed to collect and analyze data from industrial machines. This platform helps industries optimize operations, predict maintenance needs, and improve efficiency. GE's transformation into a digital industrial company has allowed it to offer innovative solutions to its clients and stay ahead in a competitive market.

4. Starbucks

Starbucks has successfully used digital transformation to enhance customer experience and streamline operations. The company developed a mobile app that allows customers to order and pay ahead, reducing wait times and improving convenience. Additionally, Starbucks leverages data analytics to personalize marketing efforts and loyalty programs. The introduction of the Starbucks Rewards program has significantly increased customer engagement and retention.

5. Nike

Nike embraced digital transformation by integrating digital technology into its products and business model. The company developed the Nike+ ecosystem, which includes wearable technology, a mobile app, and a community platform. These innovations allow users to track their fitness activities, set goals, and connect with other fitness enthusiasts. Nike also uses data analytics to optimize supply chain management and create personalized marketing campaigns, enhancing customer experience and operational efficiency.

6. Domino's Pizza

Domino's Pizza transformed its business model by heavily investing in digital technology. The company developed an online ordering system and a mobile app that allows customers to place orders easily. Domino's also introduced the "Pizza Tracker," which provides real-time updates on the status of an order. These digital initiatives have significantly improved customer experience and streamlined operations, leading to increased sales and market share.

7. Siemens

Siemens, a global industrial powerhouse, undertook a digital transformation to integrate its traditional engineering expertise with digital technologies. Siemens developed the MindSphere platform, an industrial IoT solution that connects products, plants, systems, and machines, enabling customers to harness the power of data. This platform allows for predictive maintenance, energy management, and asset optimization, providing significant value to Siemens' clients and positioning the company as a leader in the digital industrial revolution.

8. Walmart

Walmart has embraced digital transformation to compete with e-commerce giants like Amazon. The company invested in its online platform, enhancing its e-commerce capabilities and integrating its physical stores with digital operations. Walmart introduced features such as online grocery ordering with curbside pickup, mobile payment options, and personalized shopping experiences. These digital initiatives have helped Walmart maintain its position as a leading retailer by providing customers with seamless and convenient shopping experiences.

9. DBS Bank

DBS Bank, a leading financial services group in Asia, successfully transformed its business by adopting a digital-first approach. The bank invested in cloud computing, AI, and machine learning to improve its services and customer experience. DBS launched the DigiBank app, a mobile-only bank that offers seamless and convenient banking services. The bank's digital transformation efforts have led to increased customer satisfaction, operational efficiency, and market share.

10. Procter & Gamble (P&G)

Procter & Gamble implemented digital transformation strategies to enhance its supply chain, product innovation, and customer engagement. P&G utilized data analytics and IoT to optimize its manufacturing processes, reduce costs, and improve product quality. The company also developed digital marketing campaigns and personalized consumer experiences using data-driven insights. P&G's digital initiatives have enabled it to stay competitive and meet changing consumer demands.

These examples demonstrate how companies across various industries have successfully leveraged digital transformation to enhance their operations, improve customer experiences, and maintain a competitive edge. By adopting digital technologies and innovative strategies, these organizations have achieved significant growth and success in the digital era.

Table of Contents

Part 1

Digital Transformation

Chapter 1

Understanding Digital Transformation.

In the early 2000s, growing up in Nigeria, I vividly recall the excitement sparked by the arrival of Global System for Mobile Communications (GSM) technology. This innovation marked a new era in mobile communication, not only for us but also for many in developed countries. Gone were the days of scheduling appointments to wait by landline phones; now, we can receive calls anytime, anywhere.

Although initially costly and beyond the reach of many, GSM swiftly transformed our understanding of communication. It addressed the limitations of traditional landlines, leading to widespread adoption globally. As time progressed, subsequent generations (3G, 4G, and now 5G) have further evolved the GSM landscape. The mobile phones themselves have undergone stylistic and aesthetic changes to meet evolving consumer needs. There was no doubt that the introduction of GSM revolutionized the telecommunication industry.

In addition to their positive impact on Nigerians' lives, mobile phones also drive the ongoing growth of digital inclusion within the country. They empower users to engage in information exchange for both business and social interactions, enhancing productivity and facilitating greater access to information.

The widespread adoption of mobile devices has significantly bolstered the expansion of telecom services globally, extending its influence into sectors like finance, healthcare, and beyond. These devices act as crucial portals for digital and socio-economic transformation globally, as they provide essential internet access that would otherwise be inaccessible.

Today, digital transformation centers on numerous technological advancements like artificial intelligence, Big Data, and others, which are utilized to augment and enhance human work. Approximately 70% of companies in the United States and Europe employ at least one advanced digital technology on average. At its core, digital transformation involves leveraging digital technologies to create new or modify existing business processes, culture, and customer experiences to meet changing business and market requirements.

Let's explore the nine (9) key core concepts of digital transformation:

– Customer Experience (CX) Transformation

We have all heard the term, "Customers are always right." While this myth has been debunked several times by different pundits, it is still important to put customers' needs at the forefront of digital transformation. Improving customer interactions across all touchpoints using digital channels such as websites, mobile apps, social media, etc. This often involves personalization, responsiveness, and seamless integration. CX encompasses every touchpoint a customer has with a brand, from initial awareness through post-purchase support. Successful CX transformation aligns business goals with customer expectations, fostering loyalty, advocacy, and sustainable growth. It involves a commitment to innovation, agility, and a deep understanding of customer needs in an ever-evolving marketplace.

Transformation, in this context, refers to the strategic initiatives and changes a company undertakes to improve CX significantly.

1. **Operational Agility** – An organization's ability to quickly adapt and respond to environmental changes while sustaining operational efficiency and effectiveness is crucial. This requires the capacity to make prompt decisions, alter strategies, and adjust processes in response to shifting market conditions, customer demands, competitive pressures, or internal factors.

Utilizing digital tools and technologies is vital for streamlining operations, enhancing efficiency, and swiftly responding to market changes. This may involve automating processes, employing data-driven decision-making, and adopting cloud computing in the current dynamic landscape.

2. **Business Model Innovation** – Organizations reimagine traditional business models or develop new ones through digital technologies. This involves redefining and restructuring the core elements of how a company creates, delivers, and captures value.

It extends beyond product or service innovation to include fundamental changes in strategies, processes, and organizational structures. By embracing this approach, businesses can unlock new revenue streams, improve customer experiences, and strengthen their competitive advantage.

Table 1:1

Business Models
Examples of some of the most popular business models in today's world:
Subscription: Pay monthly to access a product (Office 365, Canva, Netflix & Spotify).
On-demand: Access product as part of a service for a limited time (Amazon Prime video)
Freemium: offers basic services for free and charges for premium (Dropbox, LinkedIn)
Software as a Service (SaaS): Salesforce, Google cloud platform)

3. **Cultural Shift**: Fostering a digital-first mindset across the organization, encouraging innovation, collaboration, and continuous learning. This involves breaking down silos, promoting data literacy, and embracing change.

Organizational culture is the bedrock upon which successful digital transformation is built. By fostering a culture that embraces change, drives innovation, enhances collaboration, and prioritizes agility, organizations can navigate the complexities of digital transformation more effectively.

Ultimately, a strong, adaptive culture not only supports the successful implementation of digital initiatives but also ensures their sustainability and long-term impact.

4. **Data-Driven Insights:** Harnessing the power of data analytics and AI to gain actionable insights into customer behavior, market trends, and operational performance. This enables informed decision-making and proactive strategies.

Data-driven insights are fundamental to the success of digital transformation. By harnessing the power of data, organizations can make more informed decisions, understand their customers better, optimize operations, and stay ahead of market trends. Implementing a data-driven approach requires the right tools, skilled personnel, and a supportive organizational culture, but the benefits far outweigh the challenges. Embracing data-driven insights paves the way for a more agile, efficient, and competitive organization in the digital age.

5. **Technology Integration**: Embracing emerging technologies such as IoT (Internet of Things), AI (Artificial Intelligence), blockchain, and VR/AR (Virtual Reality/Augmented Reality) to create new opportunities and improve existing processes.

Technology integration is a vital aspect of digital transformation, enabling organizations to create a cohesive and efficient digital ecosystem. By seamlessly incorporating various technologies into their operations, organizations can improve efficiency, enhance customer experiences, foster innovation, and achieve cost savings. Successful technology integration requires careful planning, the right tools, effective change management, and continuous optimization. Embracing technology integration paves the way for a more agile, competitive, and forward-thinking organization in the digital age.

6. **Cybersecurity and Risk Management**: Addressing the challenges of digital transformation by ensuring robust cybersecurity measures, compliance with regulations, and managing risks associated with digital operations.

In the era of digital transformation, cybersecurity and risk management have become paramount for organizations seeking to protect their digital assets, maintain customer trust, and ensure regulatory compliance. As cyber threats continue to evolve, staying vigilant and proactive in cybersecurity practices is essential for safeguarding the future of the organization in the digital age.

7. **Sustainability and Social Responsibility**: Incorporating environmental and social considerations into digital strategies, promoting sustainability practices, and contributing positively to society is essential for a modern organization,

By adopting sustainable practices, ethical supply chain management, community engagement, and transparency, businesses can enhance their reputation, attract top talent, and drive innovation.

A successful digital transformation integrates these interconnected elements collectively to drive growth, efficiency, and competitiveness in the digital era.

Why do Organization Transform - Innovate, and Adapt

In today's world, individuals have grown accustomed to the benefits and comfort that technology provides, enjoying the personalized digital experiences and real-time information it offers.

Organizations transform to remain competitive, meet evolving customer demands, and leverage new opportunities in an ever-changing business landscape. They seek to enhance customer experience through personalization and seamless interactions and improve operational efficiency through automation and agile processes. Digital transformation drives innovation and growth by enabling new business models and product developments. It helps organizations gain a competitive advantage, differentiate themselves, and make data-driven decisions using insights and analytics.

Additionally, transformation allows companies to adapt to market changes, comply with regulations, and manage risks. Optimizing the workforce, fostering collaboration, and supporting remote work are also key benefits. Sustainability and social responsibility objectives are met through more efficient and transparent operations. Financial imperatives, such as cost reduction and revenue generation, are also significant drivers, ensuring long-term relevance and success in the digital age.

The drivers behind digital transformation include a combination of external pressures and internal opportunities that push organizations to innovate and adapt. These drivers can be categorized into several key areas:

Customer Expectations: Encompass the need for personalized experiences tailored to individual needs, seamless and efficient interactions across digital touchpoints, and quick, effective responses to queries and issues.

Market Dynamics: Market dynamics play a very important role in influencing digital transformation. It encompasses a complex interplay of competitive pressures, technological advancements, regulatory environments, customer expectations, and economic factors—all of which shape and accelerate digital transformation across industries. Organizations that understand these dynamics and proactively adapt are better positioned to thrive in today's digital economy.

Technological Advancements: these are significant changes driving digital transformation

- **Emerging Technologies**: Innovations like AI (Artificial Intelligence), IoT (Internet of Things), and blockchain are opening new avenues for businesses to innovate and transform. AI enables automation, predictive analytics, and personalized customer experiences. IoT connects devices and enables data-driven insights for operational efficiency and new service offerings. Blockchain ensures secure and transparent transactions across various industries.

- **Data Analytics:** The ability to gather, store, and analyze vast amounts of data is transforming decision-making processes. Advanced analytics and machine learning algorithms extract valuable insights from data, enabling businesses to optimize operations, predict trends, and personalize customer interactions.

Data analytics drives efficiency improvements and competitive advantages.

- **Cloud Computing:** Cloud services provide scalable and flexible computing resources over the internet. Businesses can leverage cloud platforms to deploy applications, store data securely, and access computing power on-demand. Cloud computing reduces IT infrastructure costs, enhances collaboration, and enables rapid deployment of digital solutions, accelerating innovation and time-to-market.

These technological advancements are not only reshaping industries but also empowering businesses to streamline processes, enhance customer experiences, and stay agile in a rapidly evolving market landscape.

Operational Efficiency: This is the ability of an organization to utilize its resources effectively to produce maximum output with minimal input, thereby achieving cost savings and optimizing performance.

- **Automation**: Automating processes involves using technology to perform tasks traditionally done by humans. This approach enhances operational efficiency, cuts costs, and reduces errors. Automation ranges from simple tasks to complex workflows, freeing up human resources for more strategic activities.

- **Agility**: Businesses must swiftly adapt to changing circumstances and market dynamics to stay competitive. Agility in digital transformation refers to the ability to respond quickly and effectively to new opportunities or challenges. This includes adopting new technologies, adjusting strategies, and reorganizing operations as needed.

- **Integration**: Seamless integration of various technologies and systems is crucial for optimizing business processes. Integration ensures that different software applications, databases, and devices can communicate and work together efficiently. This streamlines operations, enhances data visibility, and improves decision-making across the organization.

- These elements—automation, agility, and integration—are fundamental in leveraging technology to drive innovation, improve productivity, and maintain competitiveness in today's fast-paced business environment.

Regulatory and Compliance Requirements: This is a significant catalyst for digital transformation across industries. These requirements compel organizations to adopt new technologies and practices to ensure adherence to legal and regulatory standards. Key drivers include:

- **Data Protection Regulations**: Laws like GDPR and CCPA necessitate robust data protection measures, prompting businesses to implement advanced cybersecurity technologies, data encryption methods, and secure data storage solutions.

- **Security Standards:** Compliance with industry-specific security standards demands the adoption of secure communication channels, authentication protocols, and vulnerability assessment tools to protect sensitive information and prevent unauthorized access.

- **Industry Regulations**: Specific regulations in sectors such as finance (e.g., PCI DSS for payment card data), healthcare (e.g., HIPAA for patient data), and others drive digital transformation initiatives. This includes deploying electronic health records (EHR) systems, implementing financial transaction monitoring technologies, and integrating compliance reporting mechanisms.

- **Digitization of Compliance Processes**: Organizations are increasingly digitizing compliance processes to enhance transparency, accuracy, and efficiency. This involves leveraging automation tools for regulatory reporting, conducting real-time monitoring of compliance metrics, and utilizing analytics to ensure ongoing adherence to regulatory requirements.

- **Global Compliance Requirements:** With businesses operating globally, compliance with international regulations and standards becomes crucial. Digital transformation facilitates cross-border data transfers, localization of compliance strategies, and adaptation to diverse regulatory frameworks.

In essence, regulatory and compliance requirements drive organizations to embrace digital transformation by adopting innovative technologies, improving operational efficiencies, and enhancing data security measures to comply with evolving legal mandates and industry standards. This transformation not only ensures regulatory compliance but also fosters resilience and competitive advantage in a dynamic business environment.

Workforce and Talent Management: focuses on developing digital skills, supporting remote work, and enhancing collaboration among employees to meet modern business demands effectively.

- **Skills Development**: Ensuring employees are equipped with digital skills necessary to leverage new technologies and innovations.

- **Remote Work**: Adapting to the trend of remote and flexible work arrangements supported by digital tools and infrastructure.

- **Collaboration**: Enhancing communication and collaboration through digital platforms and tools, facilitating teamwork and knowledge sharing across geographies.

Effectively managing workforce and talent in these areas enables organizations to adapt, innovate, and thrive in the digital age, thereby driving overall digital transformation.

Financial Imperatives: These are critical drivers of digital transformation, influencing organizations to leverage technology for strategic financial outcomes:

- **Cost Optimization**: Embracing digital solutions to streamline operations and reduce expenses, thereby enhancing overall cost efficiency and profitability.

- **Revenue Enhancement:** Introducing innovative digital products, services, and business models to expand market reach and drive revenue growth.

- **Investment in Innovation**: Allocating resources towards digital initiatives and technological advancements to foster innovation, maintain competitive advantage, and ensure long-term financial sustainability.

These drivers collectively push organizations to embark on digital transformation journeys, aiming to stay competitive, meet customer expectations, and capitalize on new opportunities in the digital age.

Table 1:2:

Examples of companies that responded to these drivers and transformed successfully:

1. Netflix: Netflix shifted from a DVD rental-by-mail service to a streaming video platform, capitalizing on internet growth and declining physical media. This move expanded its customer base and diversified revenue streams.

2. Amazon: Amazon evolved from an online bookstore to a comprehensive e-commerce platform offering a wide array of products and services. This transformation positioned Amazon as a major retail player and enabled revenue growth through platform services.

3. Apple: Apple transitioned from a computer manufacturer to a consumer electronics company, introducing iconic products like the iPhone and iPad. This transformation broadened its market reach and generated revenue from mobile devices and related services.

4. Uber: Uber transformed from a ride-hailing service into a multi-service transportation platform, including ride-hailing, food delivery, and bike rentals. This diversification expanded its customer base and created new revenue streams.

5. Airbnb: Airbnb transitioned from a room rental service to a comprehensive travel platform, including home rentals, experiences, and travel activities. This expansion diversified its offerings and enhanced its market presence.

6. Tesla: Tesla expanded from an electric car manufacturer to a sustainable energy company, offering solar panels, batteries, and electric vehicle charging infrastructure. This transformation positioned Tesla at the forefront of clean energy innovation.

7. Zoom: Zoom transformed from a video conferencing tool into a comprehensive platform for remote work, offering virtual meetings, webinars, and team collaboration solutions. This evolution met the growing demand for remote communication and collaboration tools, particularly during the pandemic.

These transformations illustrate how companies adapt to market trends, technological advancements, and customer preferences to innovate, expand, and stay competitive in their respective industries.

Source: (Rob Llewellyn, 2024)

Chapter 2

Why Digital Transformation fails.

70% of all digital transformations fail. (Kissflow Inc., 2024)

In Chapter 1, I discussed the introduction of smartphones and the adoption of GSM in Nigeria. Blackberry was a major player in this technological shift. Blackberry captured a significant market share by focusing on email services for mobile devices (e-careers, 2024), which provided them with a competitive edge over rivals like Nokia, Ericsson, and Motorola. At its peak, Blackberry boasted over 80 million users worldwide, dominating the mobile phone industry. (McKinsey & Company., 2018) However, as competitors such as Samsung and Apple began to emphasize larger touchscreen displays and continuous innovation with Android and iOS operating systems, Blackberry struggled to meet evolving customer expectations. Their incremental adjustments fell short of transforming the customer experience, ultimately leading to their decline.

In 2016, Blackberry's CEO announced the company's departure from the smartphone industry, attributing its failure to innovate and adapt to changing trends. The challenges faced by Blackberry serve as a pertinent reminder for organizations today that digital transformation and continuous innovation are critical to staying relevant in the rapidly evolving market landscape.

As organizations endeavor to thrive in the digital age, they will inevitably confront various challenges that must be effectively navigated. Digital transformation goes beyond simply incorporating new technologies into business operations or launching a new product; it necessitates a comprehensive overhaul of how organizations operate, engage with customers and deliver value.

"Even digitally savvy industries, such as high tech, media, and telecom, are struggling. Among these industries, the success rate does not exceed 26 percent. But in more traditional industries, such as oil and gas, automotive, infrastructure, and pharmaceuticals, digital transformations are even more challenging: success rates fall between 4 and 11 percent." McKinsey (Blackberry, 2024)This holistic transformation impacts every aspect of an organization, from its business model to its corporate culture, demanding substantial change management efforts; remember the core concepts we discussed in the previous chapter.

Let's explore some of the challenges that impact digital transformation.

Challenge #1 - Strategy and Vision

In the absence of a clear roadmap and vision, digital transformation efforts risk losing focus and falling short of meaningful outcomes. Developing a strategic vision enables organizations to navigate digital transformation effectively, capitalize on opportunities for growth and differentiation, and maintain long-term success in a swiftly evolving digital environment.

The vision and mission statements should encapsulate the fundamental purpose and future aspirations of the organization. According to Kotter's principles outlined in "Leading Change," effective vision and mission statements are pivotal (Jeffrey M. Hiatt, 2003). Aligning the digital transformation strategy with these statements ensures that technological initiatives are not pursued in isolation but are integrated into a comprehensive strategic framework.

It is essential for an organization's digital transformation strategy to align with its overall mission and vision. This alignment forms the bedrock of strategic coherence, optimal resource allocation, stakeholder engagement, cultural alignment, long-term sustainability, and fostering a positive external reputation.

It helps leaders and teams prioritize initiatives that are most relevant to achieving strategic goals, avoiding distractions or investments in technologies that do not align with the organization's purpose.

By achieving this alignment, organizations can amplify the impact of their digital initiatives, reinforce their overarching objectives, and secure lasting success in an increasingly digital and competitive landscape.

Table 2.1:

Characteristics of an Effective Vison Statement
Imaginable: Conveys a picture of what the future will look like.
Desirable: Appeals to the long-term interest of employees, customers, stockholders, and others who have a stake in an organization
Feasible: Comprises of realistic, attainable goals
Focused: Is clear enough to provide guidance in decision-making
Flexible: Is general enough to allow individual initiative and alternative and alternative responses in light of changing conditions
Communicable: It is easy to communicate and can be successfully explained in five minutes.
Source: Leading Change – John P. Kotter, Harvard Business Review Press, 2012. (Jeffrey M. Hiatt, 2003)

In addition to providing a clear strategy and vison, a strong leadership that champions change, supports employee development, and provides guidance through the transition is crucial. The impact of leadership on failed digital transformation initiatives can be profound and multifaceted.

By addressing these critical aspects and developing a strategic vision, organizations can navigate digital transformation effectively, capitalize on growth opportunities, differentiate themselves in the market, and achieve sustained success in today's rapidly evolving digital landscape.

Challenge #2 – Change Management and Employee Involvement

Effective communication of the vision and strategy, along with proactive change management, helps in overcoming resistance and fostering a culture of adaptability and innovation.

Ineffective change management processes can result in confusion, resistance, and, ultimately, transformation failure. The scope and pace of change are typically substantial, intensifying the complexities and underscoring the criticality of proficient change management.

Effective change management is indispensable for achieving successful digital transformation. It involves preparing, supporting, and guiding individuals and teams within the organization to embrace and adapt to changes. This process is intricate, often encountering resistance, misunderstanding, and apprehension.

Resistance to change among employees poses a significant hurdle to the successful adoption of new technologies and workflows within organizations and can manifest in various forms. Understanding the reasons behind this resistance is crucial for effectively addressing and mitigating its impact.

One of the primary reasons employees resist change is the fear that new technologies and workflows may automate or replace their current roles. This fear is often fueled by uncertainties about how their skills will align with the new requirements or whether their positions will become redundant. Employees may perceive these changes as threatening their job security, leading to apprehension and resistanceAnother common barrier to adopting new technologies is a lack of understanding about their purpose, benefits, and how they integrate into existing workflows. When employees do not comprehend the rationale behind the change or how it will enhance their work processes, they are less likely to embrace it. Involving employees in the change process by seeking their input, addressing their concerns, and empowering them to contribute to decision-making can mitigate resistance and foster a sense of ownership.

Miscommunication or inadequate training can exacerbate this lack of understanding, further contributing to resistance. Education and training initiatives are instrumental in mitigating resistance to change by equipping employees with the knowledge, skills, and confidence needed to embrace new technologies and workflows. By prioritizing comprehensive skill development, addressing knowledge gaps, offering personalized learning paths, providing continuous support, integrating with organizational culture, and measuring effectiveness, organizations can foster a culture of continuous learning and innovation that drives sustainable growth in the digital age.

Human beings are inherently creatures of habit, and many employees may resist change simply because it disrupts their familiar routines and ways of working. Transitioning to new technologies or workflows requires adapting to unfamiliar processes, tools, or systems, which can evoke discomfort and reluctance to depart from established norms.

Change often entails a shift in power dynamics or decision-making processes within an organization. Employees who feel that they are losing control over their work environment or autonomy may resist change as a means of preserving their sense of agency and influence.

The prevailing culture within an organization can significantly influence how employees perceive and respond to change. A culture that values innovation, continuous learning, and open communication is more likely to facilitate smoother transitions and reduce resistance. Conversely, a culture that is resistant to change or punitive towards experimentation may inadvertently foster greater employee resistance.

Addressing resistance to change requires proactive measures that acknowledge and respond to these underlying concerns.

By recognizing and addressing these factors, organizations can effectively manage resistance to change and pave the way for the successful adoption of new technologies and workflows, ultimately driving innovation and competitiveness in the digital age.

Amidst these challenges, a clear understanding of potential obstacles can empower organizations to chart a more efficient and prosperous course toward digital transformation. While the journey may be arduous, strategic planning, adept change management practices, and a thorough grasp of challenges can enable businesses to establish a strong foundation for sustained success in the digital era.

Challenge #3 – Communication

Digital transformation often involves significant changes in processes, tools, and organizational culture. Effective and transparent communication about the reasons for change, how it will be implemented, its expected benefits, and how it aligns with organizational goals can help alleviate uncertainty and build employee buy-in. It addresses concerns and ensures stakeholders feel involved and informed throughout the transformation journey.

How can you effect change without informing the people impacted by it? Transformation impacts not only internal teams but everyone within the organization's value chain. Recently, digital transformation has forced a shift to online channels across most industries. In the healthcare sector, we see e-health in banking, online banking, retail, and online shopping. Most of our interactions with employees and customers have moved online, and every customer and employee experience now runs through a digital platform.

As leaders, managers, and entrepreneurs, communication should be a cornerstone of your digital transformation strategies. Clear and consistent communication ensures that everyone in the organization understands the reasons behind the transformation, the steps involved, and the anticipated benefits. It helps to address any concerns, reduce resistance, and build support among employees.

By keeping all stakeholders informed, you create a sense of transparency and trust, which is crucial for successful implementation. Regular updates and feedback loops allow for continuous engagement and the ability to address issues promptly. Moreover, effective communication facilitates alignment with organizational goals, ensuring that all efforts are directed toward a common vision.

Involving employees in the communication process can also lead to valuable insights and innovations, as they are often closest to the day-to-day operations and customer interactions. This inclusive approach not only enhances the quality of the transformation but also empowers employees, making them feel valued and integral to the process.

Prioritizing communication in your digital transformation strategy helps to ensure a smooth transition, fosters a positive organizational culture, and drives the successful adoption of new technologies and processes.

Challenge #4 – Legacy Systems Integration

Older systems may not easily integrate with new digital solutions, leading to compatibility issues and delays. This challenge arises because legacy systems are often built on outdated technologies and frameworks that are not designed to interact with modern platforms. As a result, organizations may face difficulties in achieving seamless data flow and interoperability between old and new systems.

The integration process can be complex and time-consuming, requiring significant effort to bridge the technological gap. This often involves custom development, middleware solutions, and thorough testing to ensure that the systems can communicate effectively. Additionally, legacy systems may lack the scalability and flexibility needed to support new digital initiatives, further complicating the integration process.

Delays in integration can disrupt business operations, affect customer experiences, and hinder the overall progress of digital transformation efforts. Furthermore, maintaining legacy systems alongside new solutions can increase operational costs and create a fragmented IT environment, making it harder to manage and secure.

To address these challenges, organizations should develop a clear integration strategy that includes a thorough assessment of existing systems, identification of potential compatibility issues, and a roadmap for phased integration. Investing in modern integration tools, leveraging APIs, and considering gradual system upgrades or replacements can also help mitigate risks and facilitate a smoother transition.

Integrating legacy systems with new digital solutions is a critical but challenging aspect of digital transformation. Effective planning, the right tools, and a strategic approach are essential to overcome compatibility issues and ensure successful integration.

Challenge # 5 – Skill Gaps and Talent Shortages

New tech requires new skills. This challenge is particularly acute as the pace of technological advancement outstrips the rate at which the workforce can upskill. Organizations may struggle to find and retain talent proficient in cutting-edge technologies, which are critical for implementing and maintaining digital transformation initiatives. There may be a lack of expertise in new technologies (e.g., AI, IoT, data analytics) needed for digital transformation initiatives.

The skill gaps can manifest in various ways. For example, AI and machine learning require specialized knowledge in algorithms, data modeling, and programming languages such as Python and R. IoT involves understanding complex sensor networks, connectivity protocols, and data integration. Data analytics demands expertise in big data platforms, statistical analysis, and data visualization tools. Without these skills, organizations may find it challenging to fully leverage the potential of these technologies, leading to suboptimal outcomes.

This shortage of skilled professionals can lead to several issues. Projects may be delayed or executed poorly, impacting their effectiveness and ROI. Additionally, the burden on existing staff can increase, leading to burnout and turnover, which further exacerbates the talent shortage. The inability to implement new technologies efficiently can also hinder an organization's competitive edge and slow down its overall digital transformation journey.

To address these challenges, organizations need to adopt a multifaceted approach. Investing in continuous learning and development programs can help bridge skill gaps by upskilling existing employees. Partnering with educational institutions, offering internships, and creating apprenticeship programs can also build a pipeline of future talent. Additionally, leveraging external expertise through consultants or outsourcing certain functions can provide immediate relief while long-term strategies are put in place.

Another critical strategy is to foster a culture of innovation and learning within the organization. Encouraging employees to experiment with new technologies and providing them with the resources and time to do so can help cultivate in-house expertise. Moreover, creating clear career paths and growth opportunities in emerging tech fields can attract top talent and improve retention rates.

Skill gaps and talent shortages are significant barriers to digital transformation. By investing in training and development, fostering a culture of continuous learning, and leveraging external resources, organizations can overcome these challenges and build a workforce capable of driving successful digital transformation initiatives.

Challenge # 6 – Culture and Organizational Alignment

During the pandemic, many organizations were compelled to transition to remote work. Some of these organizations, which had previously been reluctant to allow remote work, struggled to adapt quickly to this change. Their slow response in adjusting their organizational culture to be more flexible led to the loss of talented employees to competitors whose cultures were more agile and responsive.

Conversely, a few organizations had already fostered a proactive culture and organizational alignment that included remote work as a norm. They foresaw the potential need for employees to work from home and were better prepared to face the disruption caused by the pandemic.

In today's environment of continuous digital transformation, organizational culture has become increasingly crucial. Organizational culture reflects the norms, attitudes, and behaviors that collectively define an organization. It shapes the identity of the organization, guides decision-making, and influences how the organization responds to challenges, both internal and external. Imagine culture as the "PERSONA" of an organization. It plays a crucial role in shaping the work environment, fostering employee engagement, and determining overall effectiveness.

The development of culture within organizations and the decisions made by leaders play a central role in establishing and preserving a desirable culture. Building such a culture is a transformative journey that cannot be enforced solely through directives from leadership.

A culture that resists innovation, collaboration, or change can hinder the adoption and implementation of new technologies and processes. This resistance ultimately leads to difficulties in adapting to transformation.

While creating a suitable culture cannot be achieved overnight, organizations can create an environment that fosters a lasting cultural framework. As a leader, addressing cultural and organizational alignment requires a comprehensive approach that includes commitment from leadership, nurturing a supportive culture, breaking down silos, aligning incentives, and empowering employees through effective communication and skills development.

Those who successfully navigate these challenges are better positioned to achieve sustainable digital transformation and long-term success.

Challenge #7 - Data Management and Security Concerns

Managing large volumes of data securely and ensuring compliance with data protection regulations can be challenging. As organizations increasingly rely on data to drive digital transformation, the volume, variety, and velocity of data grow exponentially. This surge in data can strain existing data management systems, leading to potential inefficiencies and vulnerabilities.

Effective data management involves not only storing and processing data but also ensuring its accuracy, consistency, and accessibility. Organizations must implement robust data governance frameworks to oversee data quality, usage, and lifecycle management. This includes establishing clear policies for data classification, retention, and disposal to maintain data integrity and relevance.

Security is a critical aspect of data management. With the rise in cyber threats, organizations face the constant risk of data breaches, ransomware attacks, and other malicious activities. Protecting sensitive information requires a multi-layered security approach that includes encryption, access controls, regular audits, and continuous monitoring. Additionally, employee training on data security best practices is essential to minimize human error and insider threats.

Compliance with data protection regulations, such as GDPR, CCPA, and HIPAA, adds another layer of complexity. These regulations impose strict requirements on how data is collected, processed, stored, and shared. Non-compliance can result in hefty fines, legal consequences, and reputational damage. Organizations must stay informed about evolving regulatory landscapes and ensure their data practices align with legal standards.

Achieving compliance involves conducting regular assessments and audits to identify and address potential gaps. It also requires maintaining detailed records of data processing activities and implementing mechanisms for data subject rights, such as the right to access, rectify, and delete personal data. Automated compliance tools and legal expertise can assist in navigating these regulatory challenges.

Data management and security concerns are further compounded by the increasing use of cloud services and third-party vendors. While cloud solutions offer scalability and flexibility, they also introduce risks related to data privacy and control. Organizations must carefully evaluate the security measures of their cloud providers and ensure contractual agreements include provisions for data protection and compliance.

Managing large volumes of data securely and ensuring compliance with data protection regulations is a multifaceted challenge. Organizations must implement robust data governance frameworks, adopt comprehensive security measures, and stay abreast of regulatory changes to mitigate risks and protect their data assets effectively.

Challenge # 8 - Budget and Resource Constraints.

In Nigeria, there's a saying that attributes the soup's deliciousness to the power of wealth, suggesting that great outcomes often happen at considerable costs.

Digital transformation Initiatives often require substantial investment in new technologies, infrastructure, and skills development. Insufficient funding and resources allocated to digital initiatives can limit their scope and effectiveness. Without adequate funding, organizations may struggle to implement the necessary changes, leading to incomplete or suboptimal transformations.

The high expenses associated with advanced technologies such as AI, machine learning, IoT, and big data analytics might compel organizations to compromise either on quality or scope. Moreover, the upgrade of IT infrastructure to support these new digital tools and platforms could face delays due to inadequate resources, creating a disconnect between new capabilities and existing infrastructure.

Budgetary constraints may also hinder the organization's ability to attract and retain skilled professionals, leading to skill gaps and talent shortages. Limited financial resources might necessitate scaling back or prioritizing specific projects, resulting in a piecemeal approach that diminishes the benefits of a fully integrated digital ecosystem. Furthermore, constrained resources can stifle innovation and experimentation, preventing organizations from discovering effective solutions and achieving significant advancements.

To address budget and resource constraints, organizations should consider several strategies:

- **Prioritize Initiatives**: Focus on high-impact projects that align with strategic goals and deliver quick wins to build momentum and demonstrate value.

- **Leverage Cost-Effective Solutions**: Explore cost-effective technologies, such as open-source software, cloud services, and SaaS (Software as a Service) models, which can reduce upfront investment and provide scalability.

- **Seek External Funding**: Explore external funding options, such as grants, loans, and partnerships, to supplement internal budgets and expand resource availability.

- **Optimize Resource Allocation**: Regularly review and optimize resource allocation to ensure that funds and efforts are directed toward the most critical and value-generating activities.

- **Foster a Culture of Innovation**: Encourage a culture of innovation and continuous improvement, where employees are empowered to find creative solutions and efficiencies within existing constraints.

Achieving Digital Transformation hinges on grasping its fundamental concepts, recognizing prevalent challenges, and addressing them through robust leadership, clear communication, employee development, strategic planning, and a commitment to adapt and innovate. By prioritizing projects, utilizing cost-effective solutions, exploring external funding, optimizing resource allocation, and nurturing an innovative culture, organizations can effectively navigate obstacles and succeed in their digital initiatives.

Part 2:

Strategies and Technologies

Chapter 3

The Maze of Choices: Technologies and Strategies

In today's environment, digital transformation is driven by several key technologies that fundamentally reshape how businesses operate and interact with customers. A convergence of Cloud Computing, AI, Blockchain, Machine Learning, Big Data, Augmented Reality (AR), Virtual Reality (VR), and 5G technology is at the forefront of this transformation. These technologies collectively enable organizations to innovate, streamline processes, enhance decision-making, and deliver more personalized customer experiences, marking a significant shift in the way industries function and evolve.

Organizations must embrace change; these technologies empower them to innovate, streamline operations, and enhance customer experiences in the digital era. Strategic integration of these technologies is crucial for maintaining competitiveness and fostering sustainable growth in today's dynamic business landscape.

Artificial Intelligence (AI) stands out as the current disruptor in technology. Every organization is racing to implement AI in various operational contexts. Its impact will be pivotal across industries, including retail, finance, healthcare, and government agencies.

As a digital transformation consultant, I often find myself at the outset of an initiative faced with abundant choices. Numerous pathways promise to achieve our goals, and a plethora of vendors tout the greatness of their solutions and the transformative benefits they offer to organizations. I recognize that I am not alone in this dilemma. Despite crafting several strong business cases, some of these initiatives have not yielded the desired outcomes.

Throughout my career and through discussions with other consultants, I've observed that beyond the challenges discussed in previous chapters, a significant issue is the absence of a structured approach to determining which transformation efforts organizations should undertake. With technologies like Cloud, IoT, AI, and machine learning disrupting traditional organizational processes, it's natural for organizations to seek to integrate these solutions for competitive advantage. However, hastily made decisions often lead to unsuccessful transformations. As Yogi Berra famously said, "If you don't know where you are going, you'll end up someplace else."

To ensure that transformation efforts align with strategic goals, feasibility, and potential impact, it's crucial to follow a structured approach, highlighted in Figure 3.1. It summarizes the systematic steps that will help you identify the most suitable transformation effort.

Figure 3.1: The 10 key methods to identifying suitable transformation effort

1. **Define Strategic Objectives:** Clearly articulate the overarching goals and objectives of the transformation. These could include improving operational efficiency, enhancing customer experience, entering new markets, or adapting to technological advancements.

↓

2. **Assess Current State:** Conduct a thorough assessment of the current state of your organization. This includes evaluating existing processes, technologies, capabilities, and organizational culture.

↓

3. **Identify Transformation Opportunities:** Identify potential areas for transformation that align with your strategic objectives. These could involve adopting new technologies (like AI or IoT), optimizing processes, improving customer engagement, or restructuring organizational units.

↓

4. **Evaluate Feasibility:** Assess the feasibility of each transformation effort. Consider factors such as resource availability (financial, human, technological), potential risks, regulatory requirements, and timeline constraints.

↓

5. **Prioritize Initiatives:** Prioritize transformation initiatives based on their potential impact on achieving strategic objectives, feasibility, and alignment with the organization's capabilities and culture.

↓

6. **Develop Transformation Roadmap:** Create a detailed transformation roadmap that outlines the sequence of initiatives, timelines, resource allocation, and key milestones. This roadmap should be flexible enough to accommodate adjustments based on feedback and changing circumstances.

↓

7. **Engage Stakeholders:** Engage key stakeholders throughout the process to ensure alignment, obtain buy-in, and leverage their expertise and insights.

↓

8. **Monitor and Measure Progress:** Establish metrics and key performance indicators (KPIs) to monitor the progress and success of each transformation effort. Regularly assess and adjust strategies based on these metrics.

↓

9. **Iterate and Improve:** Transformation is an iterative process. Continuously solicit feedback, learn from successes and failures, and refine your approach to maximize the impact of the transformation efforts.

↓

10. **Communicate and Celebrate Successes:** Effectively communicate milestones and successes achieved through transformation efforts to build momentum, maintain morale, and reinforce the value of ongoing initiatives.

The steps outlined above are not universally standard and may differ from one leader or author to another. Nevertheless, by adhering to these steps, you can methodically identify and implement the most suitable transformation effort for your organization.

- *Strategic Objectives*

Strategic objectives are crucial for guiding digital transformation initiatives and ensuring they align with the overarching goals of the organization. Here are common strategic objectives that organizations may focus on during digital transformation:

One of the primary objectives is to ***enhance customer experience***. This involves improving customer satisfaction, loyalty, and engagement through better services and interactions. Key actions to achieve this include implementing customer relationship management (CRM) systems to manage customer interactions effectively, personalizing customer experiences using data analytics, and enhancing customer support with AI chatbots and self-service portals. These efforts aim to create seamless and personalized customer journeys, ultimately leading to higher satisfaction and retention rates.

Another critical objective is to ***increase operational efficiency***. Organizations strive to streamline operations to reduce costs, improve productivity, and optimize resource utilization. This can be achieved by automating repetitive and manual processes, implementing enterprise resource planning (ERP) systems to integrate and manage core business processes, and utilizing predictive maintenance and IoT for equipment and asset management. By doing so, organizations can achieve more efficient and cost-effective operations, freeing up resources for more strategic activities.

Driving revenue growth is also a key strategic objective. This involves boosting revenue by entering new markets, expanding product lines, or improving sales strategies. Actions to support this objective include developing e-commerce platforms to reach new customer segments, using data analytics for targeted marketing campaigns, and introducing new digital products or services. These initiatives help organizations tap into new revenue streams and enhance their market presence.

Fostering innovation and agility is essential for staying competitive in a rapidly changing market. Organizations aim to promote a culture of innovation and agility to quickly adapt to market changes and emerging technologies. Encouraging experimentation and piloting new technologies, implementing agile methodologies in project management, and establishing innovation labs or incubators are key actions that support this objective. This approach enables organizations to stay ahead of the curve and continuously innovate.

Enhancing data-driven decision-making is another strategic objective. Leveraging data analytics to make informed decisions and gain competitive insights is crucial. Implementing business intelligence (BI) and analytics tools, developing a centralized data repository or data lake, and training staff on data literacy and analytics skills are important steps to achieve this. By doing so, organizations can harness the power of data to drive strategic decisions and gain a competitive edge.

Improving cybersecurity and compliance is critical for protecting data and maintaining trust. Organizations aim to strengthen cybersecurity measures and ensure compliance with regulations. This involves implementing advanced cybersecurity protocols and tools, conducting regular security audits and risk assessments, and ensuring compliance with industry regulations such as GDPR and HIPAA. These actions help safeguard sensitive information and mitigate risks.

Enhancing employee productivity and engagement is also a strategic objective. Improving employee productivity and engagement through better tools, processes, and work environments is vital. Providing digital collaboration tools such as Slack and Microsoft Teams, offering continuous learning and development opportunities, and fostering a flexible and remote work environment are key actions to achieve this. These initiatives help create a motivated and productive workforce.

Expanding market reach and penetration is another objective. Organizations aim to increase market reach and penetration through digital channels and global expansion. Optimizing digital marketing strategies, expanding into new geographic markets using digital platforms, and partnering with global distributors or e-commerce platforms are actions that support this objective. These efforts help organizations reach new customers and grow their market share.

In summary, defining clear strategic objectives such as enhancing customer experience, increasing operational efficiency, driving revenue growth, fostering innovation, enhancing data-driven decision-making, improving cybersecurity, boosting employee productivity, and expanding market reach is essential for guiding digital transformation initiatives. Aligning digital initiatives with these objectives, setting KPIs, and developing a detailed roadmap ensures that transformation efforts are directed towards achieving meaningful and measurable outcomes, driving overall business success.

- ### *Access Current State*

Assessing the current state of your organization is a critical first step in any digital transformation journey. This involves conducting a comprehensive evaluation of existing processes, technologies, capabilities, and organizational culture. Begin by mapping out current workflows and identifying inefficiencies or bottlenecks that hinder productivity. Examine the technology stack to determine if current systems are outdated or lack the integration needed to support new digital initiatives.

Evaluate the skillsets and capabilities of your workforce to identify gaps that may need to be addressed through training or hiring. Additionally, assess the organizational culture to understand how it might support or resist change; a culture that is adaptable and open to innovation is crucial for successful digital transformation. This thorough assessment provides a clear understanding of the starting point, helping to identify areas for improvement and setting a solid foundation for developing a targeted and effective digital transformation strategy.

- ### *Identify Transformation Opportunities*

Identifying transformation opportunities is a crucial phase that involves pinpointing specific areas within your organization that can benefit from change and innovation aligned with your strategic objectives. Start by exploring how adopting new technologies such as artificial intelligence (AI), the Internet of Things (IoT), or cloud computing can enhance operational efficiency, data analytics, and decision-making processes.

Look for opportunities to optimize existing workflows by automating repetitive tasks, thus freeing up valuable human resources for more strategic activities. Improving customer engagement is another critical area where digital tools can provide personalized experiences, streamline customer service, and enhance satisfaction.

Additionally, organizational restructuring should be considered to foster a more agile and collaborative environment that can quickly adapt to market changes and technological advancements. This holistic approach ensures that identified opportunities not only support immediate needs but also contribute to long-term strategic goals, driving sustainable growth and competitive advantage. We will delve more into this in the next chapter.

- *Evaluate Feasibility*

Evaluating the feasibility of each digital transformation effort is essential to ensure that the initiatives are practical and achievable within the organizational context. This process involves a detailed analysis of several key factors.

First, assess resource availability, including financial resources to fund the initiatives, human resources with the necessary skills and expertise, and technological resources that can support the proposed changes. It's crucial to determine whether the organization has the capacity to undertake the transformation or if additional investments or partnerships are needed. Potential risks should be identified and analyzed, including operational, financial, and reputational risks, along with mitigation strategies to address them.

Regulatory requirements must be carefully considered to ensure compliance with industry standards and legal obligations, which can impact the scope and implementation of transformation efforts.

Lastly, evaluate timeline constraints by mapping out a realistic schedule for each initiative, ensuring that deadlines are achievable and that the project does not disrupt ongoing operations. This comprehensive feasibility assessment helps prioritize initiatives that are not only aligned with strategic goals but are also realistically achievable, setting the stage for successful execution and sustainable transformation.

- *Prioritize Initiatives*

Prioritizing initiatives is a critical step in ensuring the success of a digital transformation strategy, as it allows the organization to focus on the most impactful projects first. Begin by aligning each initiative with the organization's strategic goals to ensure that they contribute meaningfully to long-term objectives such as market expansion, customer satisfaction, or operational efficiency.

Use a scoring model to evaluate and rank initiatives based on various criteria, including their potential value, the effort required for implementation, and the anticipated return on investment. Consider the urgency of each initiative, prioritizing those with time-sensitive benefits or those that address immediate pain points.

Additionally, assess the feasibility of each initiative and consider resource availability, potential risks, and regulatory compliance. Engage with key stakeholders to gather insights and build consensus on priorities, ensuring that there is broad support for the selected initiatives. By systematically evaluating and ranking initiatives, the organization can allocate resources effectively, manage risks, and drive significant, sustainable improvements in alignment with its digital transformation goals.

- *Develop Transformation Roadmap*

Developing a transformation roadmap is a crucial step in guiding the successful execution of digital transformation initiatives. This roadmap serves as a strategic blueprint that outlines the sequence of activities, milestones, and timelines necessary to achieve the desired transformation outcomes. Begin by defining clear objectives and goals for the transformation, ensuring they are aligned with the overall strategic vision of the organization. Break down these goals into specific, actionable initiatives, detailing the scope, objectives, and expected outcomes of each.

Next, establish a timeline that sequences the initiatives logically, considering dependencies and resource availability. Assign realistic deadlines to each phase of the project, ensuring there is adequate time for planning, execution, and review. It's important to incorporate flexibility within the roadmap to adapt to unforeseen challenges or opportunities that may arise.

Resource planning is another critical component; identify the financial, human, and technological resources required for each initiative and allocate them appropriately. Develop a budget that reflects the cost of implementation and potential returns, ensuring financial feasibility and sustainability.

Engage key stakeholders throughout the development of the roadmap to gather input, foster buy-in, and ensure alignment across the organization. This collaborative approach helps to address concerns and leverage insights from different perspectives, enhancing the robustness of the roadmap.

Include key performance indicators (KPIs) and metrics to monitor progress and measure the success of each initiative. Establish regular review and feedback loops to assess progress against the roadmap, allowing for adjustments and realignments as necessary.

Finally, communicate the roadmap clearly and transparently to all stakeholders, ensuring everyone understands their roles, responsibilities, and the overall vision. A well-developed transformation roadmap not only provides a clear path forward but also helps to manage expectations, mitigate risks, and drive cohesive, coordinated efforts toward achieving the digital transformation objectives.

- *Engage Stakeholders*

Engaging stakeholders is a pivotal component of any successful digital transformation initiative. It ensures that the transformation efforts have broad support, is well-aligned with organizational needs, and can overcome potential resistance. Begin by identifying all relevant stakeholders, including executives, managers, employees, customers, suppliers, and partners. Understanding the interests, concerns, and influence of each stakeholder group is crucial for tailoring engagement strategies.

Start with executive leadership to secure buy-in and sponsorship, as their support is vital for resource allocation and strategic alignment. Clearly communicate the vision, objectives, and benefits of the digital transformation to this group, emphasizing how it aligns with the organization's strategic goals. Use data and case studies to build a compelling case for the proposed changes.

Middle managers and frontline employees are often the ones most directly affected by transformation initiatives. Engage them early in the process through workshops, focus groups, and feedback sessions to gather their insights and address their concerns. This inclusive approach helps identify practical challenges and solutions for those who understand day-to-day operations best. Additionally, involving employees fosters a sense of ownership and reduces resistance to change.

For customers and partners, communication should focus on how the transformation will enhance their experience and bring value. Use surveys, interviews, and advisory boards to solicit their input and ensure their needs are being considered. Transparent and continuous communication about the progress and benefits of the transformation helps build trust and maintain strong relationships.

Internally, cross-functional teams that include representatives from different departments should be established to collaborate on planning and implementation. These teams can act as change champions, promoting the transformation within their respective areas and ensuring alignment across the organization.

Regularly update all stakeholders on the progress of the transformation through newsletters, meetings, and dashboards. Celebrate milestones and successes to maintain momentum and positive engagement. Provide training and support to help stakeholders adapt to new processes and technologies, ensuring a smooth transition.

Incorporate feedback mechanisms to continuously improve the engagement process. Actively listen to stakeholder concerns and suggestions and adjust as needed to address any issues that arise. This iterative approach helps refine strategies and ensure that the transformation is not only successful but also sustainable in the long term.

In summary, effective stakeholder engagement involves clear communication, inclusive participation, continuous feedback, and proactive support. By aligning stakeholder interests with transformation goals and maintaining open, transparent communication, organizations can build the necessary support and collaboration to drive successful digital transformation.

- *Monitor and Measure Progress*

Monitoring and measuring progress is a critical aspect of managing a digital transformation initiative, ensuring that the organization stays on track toward achieving its strategic objectives. This process involves establishing key performance indicators (KPIs), setting up robust tracking systems, and implementing regular review mechanisms to assess the effectiveness of the transformation efforts.

Begin by identifying clear and measurable KPIs that align with the strategic goals of the digital transformation. These indicators should cover various aspects such as operational efficiency, customer satisfaction, revenue growth, and employee engagement. KPIs might include metrics like time savings from automated processes, customer retention rates, digital adoption rates among employees, and return on investment (ROI) for specific projects. Each KPI should have a defined target that represents success.

Set up tracking systems that can accurately collect and report data related to the chosen KPIs. This might involve integrating advanced analytics platforms, project management tools, and dashboard solutions that provide real-time visibility into the progress of each initiative. Leveraging technologies such as AI and machine learning can enhance the accuracy and predictive capabilities of these tracking systems, enabling more informed decision-making.

Establish a schedule for regular review meetings where the progress of the digital transformation is assessed against the established KPIs. These meetings should involve key stakeholders and project leaders to ensure a comprehensive evaluation. During these reviews, analyze the collected data to identify trends, achievements, and areas needing improvement. Use these insights to make data-driven decisions and adjustments to the transformation strategy as necessary.

As part of the monitoring process, proactively identify any challenges or roadblocks that might impede progress. This could include issues related to resource allocation, technological barriers, or resistance to change within the organization. Develop mitigation strategies to address these challenges promptly, ensuring that they do not derail the overall transformation efforts.

Adopt a mindset of continuous improvement by using the insights gained from monitoring and measuring progress to refine and enhance the transformation initiatives. Encourage a culture of feedback where employees and stakeholders can provide input on what is working well and what needs adjustment. This iterative approach helps fine-tune processes, technologies, and strategies to better align with evolving goals and market conditions.

Maintain transparency by regularly communicating progress to all stakeholders. Use visual tools like dashboards, reports, and presentations to share updates on KPIs and overall transformation milestones. Celebrating successes and acknowledging contributions can help maintain momentum and foster a positive attitude toward the transformation journey.

Compare your progress against industry benchmarks and best practices to ensure that the transformation efforts are competitive and aligned with leading standards. This external perspective can provide valuable insights and opportunities for further improvement.

Incorporate scenario planning into your monitoring strategy to anticipate potential future challenges and opportunities. By considering various scenarios, the organization can develop contingency plans and be better prepared to adapt to changes in the business environment.

Periodically review and realign the strategic goals of the digital transformation to ensure they remain relevant in a dynamic market. This might involve updating KPIs, shifting focus to new priorities, or accelerating certain initiatives based on market trends and organizational needs.

By systematically monitoring and measuring progress, organizations can ensure that their digital transformation initiatives remain on course, deliver expected benefits, and continuously evolve to meet changing demands. This disciplined approach not only enhances accountability and transparency but also drives sustained success in the digital era.

- *Iterate and Improve*

Iterating and improving is a fundamental principle in the successful execution of digital transformation initiatives. This approach ensures that the transformation is not a one-time event but a continuous process of enhancement and adaptation. By fostering a culture of iteration and continuous improvement, organizations can remain agile, responsive, and competitive in a rapidly changing digital landscape.

Creating an effective feedback loop is the cornerstone of iteration and improvement. This involves regularly collecting input from various stakeholders, including employees, customers, and partners. Use surveys, interviews, and feedback forms to gather insights on what is working well and where improvements are needed. Ensure that this feedback is systematically reviewed and integrated into the transformation process.

Before full-scale implementation, pilot new initiatives or changes on a smaller scale. This allows for testing in a controlled environment and helps identify potential issues without significant risk. Collect data from these pilot tests to evaluate the effectiveness and feasibility of the initiatives. Use these insights to refine and improve the approach before broader deployment.

Adopting agile methodologies that emphasize iterative development and continuous feedback can significantly enhance the digital transformation process. Agile practices, such as sprint planning, daily stand-ups, and retrospective meetings, promote regular reassessment and adjustment of initiatives. This framework allows teams to quickly respond to changes, integrate new insights, and continuously improve processes and outcomes.

Consistently measuring the impact of implemented changes using key performance indicators (KPIs) and other relevant metrics is crucial. Analyze the data to determine whether the changes are delivering the expected benefits and meeting strategic objectives. If the results fall short, use this information to understand why and identify areas for improvement.

Encouraging a culture of innovation where employees are empowered to experiment and suggest new ideas is vital. Create safe spaces for experimentation where failure is seen as a learning opportunity rather than a setback. Recognize and reward innovative thinking and successful implementations to motivate continuous improvement efforts across the organization.

Investing in continuous learning and development so that employees can keep up with the latest technologies, methodologies, and industry trends is essential. Provide training programs, workshops, and access to online courses that enhance skills and knowledge. A well-informed and skilled workforce is better equipped to contribute to ongoing improvement efforts and drive successful transformation.

Regularly reviewing and updating business processes ensures they remain efficient and aligned with digital transformation goals. Conduct process audits to identify inefficiencies, redundancies, and areas for optimization. Implement process improvements incrementally and monitor their impact to ensure they deliver the desired results.

Leveraging advanced technologies such as artificial intelligence, machine learning, and data analytics can gain deeper insights and drive improvements. These technologies can help identify patterns, predict outcomes, and automate decision-making processes, leading to more informed and effective improvements.

Maintaining clear and transparent communication about changes and progress throughout the organization is crucial. Keep stakeholders informed about what is being iterated, why changes are being made, and how these changes contribute to overall goals. Transparency builds trust and ensures that everyone is aligned and committed to the continuous improvement journey.

Regularly benchmarking performance against industry standards and competitors can identify areas for improvement. Use these benchmarks to set higher goals and challenge the organization to strive for excellence. Comparing against best practices can reveal gaps and opportunities that might not be apparent from an internal perspective.

Finally, be prepared to adapt and evolve strategies based on feedback, data, and changing market conditions. Flexibility is key to staying relevant and competitive. Continuously refine and adjust the digital transformation roadmap to reflect new insights, emerging technologies, and shifting priorities.

By embracing iteration and continuous improvement, organizations can ensure that their digital transformation initiatives remain dynamic, effective, and aligned with long-term goals. This proactive approach fosters a culture of excellence, drives sustained success, and positions the organization to thrive in the digital age.

- *Communicate and Celebrate Successes*

Communicating and celebrating successes is vital to maintaining momentum and morale during digital transformation initiatives. Effective communication ensures that all stakeholders are aware of the progress being made, the achievements attained, and the benefits realized from these efforts. Regular updates through various channels such as newsletters, emails, meetings, and internal social platforms help keep everyone informed and engaged.

Transparency in sharing both the successes and the lessons learned fosters a culture of openness and continuous improvement. Celebrating successes, whether big or small, plays a crucial role in motivating teams and reinforcing the value of their hard work. Recognizing individual and team contributions through awards, public acknowledgments, or special events not only boosts morale but also encourages a sense of ownership and pride in the transformation journey. Highlighting success stories and case studies can inspire others within the organization to embrace change and strive for excellence. Additionally, celebrating milestones provides an opportunity to reflect on the progress made, reassess goals, and realign strategies if necessary.

This practice helps maintain a positive and forward-looking atmosphere, ensuring sustained enthusiasm and commitment towards achieving the digital transformation objectives. Communicating and celebrating successes effectively builds a resilient organizational culture that values achievement and continuous improvement, driving long-term success in the digital era.

By embracing these steps, organizations and leaders can ensure their digital transformation initiatives remain dynamic, effective, and aligned with long-term goals, driving sustained success in the digital era.

In the next chapter, I will explore these steps in greater detail to create a framework for refining strategies, ensuring they align with strategic objectives and maximize long-term success.

Chapter 4

Filtering Strategies for Success

Let's face it: navigating the dynamic and intricate landscape of digital transformation presents organizations with numerous options, technologies, and initiatives. The sheer abundance of available technologies, initiatives, and options can quickly become overwhelming, leading to confusion, scattered efforts, and, ultimately, less-than-optimal outcomes.

This is where filtering strategies come into play, offering a structured method to sift through these possibilities and focus on what truly matters for the organization's success. To address these challenges and ensure a focused and effective transformation journey, organizations rely on filtering strategies.

These strategies involve a systematic approach designed to evaluate, prioritize, and select initiatives that are most relevant and impactful. The goal is to ensure that the organization's efforts are concentrated on areas that will yield the highest return on investment and best align with its goals, resources, and capabilities.

A well-constructed filtering framework is crucial for guiding an organization through the complexities of digital transformation. By implementing a filtering framework, organizations can streamline their efforts, optimize resource allocation, and achieve coherent and successful transformation outcomes. Key considerations include aligning initiatives with overarching objectives to establish a clear purpose and direction. Stakeholder analysis helps understand diverse needs and expectations, ensuring initiatives resonate across relevant parties.

Conducting a cost-benefit analysis enables the evaluation of feasibility and potential return on investment, prioritizing projects with significant value. Assessing risks identifies potential challenges and dependencies, informing strategic decision-making. Evaluation of resource availability ensures adequate support for chosen initiatives while prioritizing scalability and flexibility, which enables adaptation to evolving circumstances.

Let's explore these flittering strategies further:

Digital Transformation Filtering Framework

This framework provides a structured approach to prioritizing digital transformation initiatives, ensuring alignment with strategic goals, and maximizing impact. It consists of several key components, each with specific steps and considerations.

Figure 4.1: Digital Transformation Filtering Framework

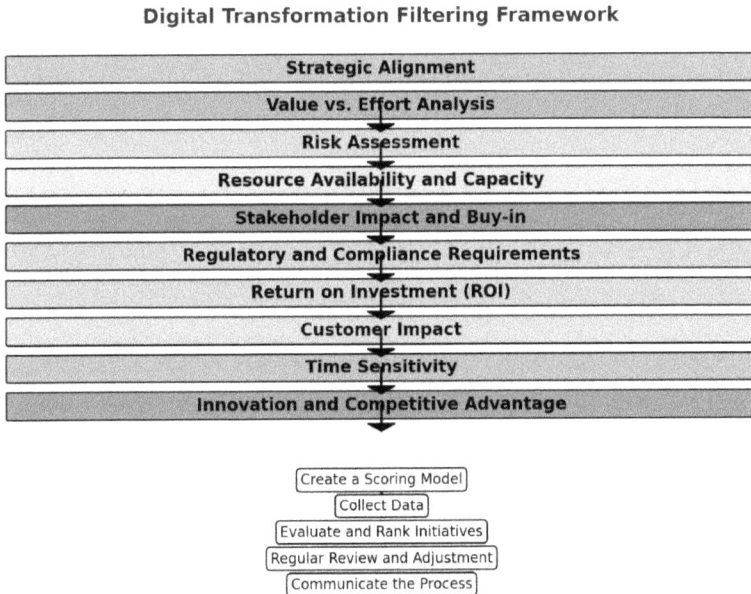

Digital Transformation Filtering Framework

Strategic Alignment
Value vs. Effort Analysis
Risk Assessment
Resource Availability and Capacity
Stakeholder Impact and Buy-in
Regulatory and Compliance Requirements
Return on Investment (ROI)
Customer Impact
Time Sensitivity
Innovation and Competitive Advantage

Create a Scoring Model
Collect Data
Evaluate and Rank Initiatives
Regular Review and Adjustment
Communicate the Process

This framework visually outlines the key steps and criteria to prioritize digital transformation initiatives. The flow from strategic alignment through to innovation and competitive advantage is shown, with arrows indicating the progression through the process. Implementation steps are displayed at the bottom, illustrating the final phase of evaluating and refining initiatives.

If your goal is to build and lead an adaptive organization that is proactive in its digital transformation efforts to survive market uncertainties, you, as a leader, must ensure you fully understand and implement a comprehensive framework. This framework not only guides the evaluation and selection of initiatives but also fosters a culture of continuous improvement, agility, and resilience.

Understanding and Implementing the Framework

Strategic Alignment ensures initiatives align with organizational goals.

The common denominator to most successful transformation efforts, if not all, is an alignment with the organization's strategic goals and objectives. Not surprisingly, a successful digital transformation requires a well-thought-out and integrated strategy. The strategy should include a clear vison, backed by strategic imperatives and quantifiable business outcomes.

To ensure alignment, transformation leaders need to develop a clear digital transformation roadmap aligned with strategic goals and evaluate each initiative to determine its alignment with this roadmap, assigning higher priority to initiatives that directly support strategic objectives such as enhancing customer experience, operational efficiency, or entering new markets.

Steps:

a. Develop a digital transformation roadmap aligned with strategic objectives.

b. Evaluate initiatives for their contribution to these goals.

c. Prioritize initiatives that enhance customer experience, operational efficiency, or market expansion.

Figure 4.2: Examples of Alignment with Digital Initiatives

Strategic Objective	Digital Initiative Example
Enhance Customer Experience	Implementing a CRM system to manage customer interactions and improve satisfaction.
Increase Operational Efficiency	Automating supply chain processes to reduce costs and improve accuracy.
Drive Revenue Growth	Developing an e-commerce platform to reach new customer segments.
Foster Innovation and Agility	Establishing an innovation lab to experiment with new technologies.
Enhance Data-Driven Decision Making	Implementing a BI tool to analyze market trends and support strategic decisions.
Improve Cybersecurity and Compliance	Deploying advanced threat detection systems to protect against cyber threats.
Enhance Employee Productivity and Engagement	Providing digital collaboration tools and remote work options.
Expand Market Reach and Penetration	Launching a global digital marketing campaign to increase brand awareness and sales.

Value vs. Effort Analysis: Assess the value and effort of initiatives.

Value vs. Effort Analysis is a strategic tool that is used to prioritize digital transformation initiatives by comparing the potential value they deliver against the effort required to implement them. This approach helps organizations focus on initiatives that offer the highest returns with the least amount of resources, ensuring efficient allocation of time, budget, and workforce.

Steps:

a. Create a 2x2 matrix with value and effort as axes.

b. Plot each initiative on the matrix.

c.   Prioritize high-value, low-effort initiatives for quick wins and high-value, high-effort initiatives for strategic impact.

Conducting a Value vs. Effort Analysis allows organizations to strategically prioritize digital transformation initiatives based on their potential impact and the resources required. This systematic approach ensures that efforts are focused on projects that offer the greatest benefits with manageable effort, aligning with overall strategic goals and optimizing resource utilization. As a leader, understanding and implementing this analysis will enable you to guide your organization through a successful and efficient digital transformation journey.

The Importance of Risk Assessment in Digital Transformation.

As someone who thrives on adrenaline, I believe life without risk is boring. However, for those who are risk-averse, avoiding risk may be their optimal state of being. Innovation is inherently risky, and if you embrace this mindset, you are ready to tackle the challenges that come with it. While innovation and risk-taking are powerful drivers of entrepreneurial success, they also require a delicate balance. Innovating without considering potential risks can lead to reckless decisions, whereas excessive risk aversion may stifle creativity and growth.

Risk assessment is a critical component of digital transformation planning. It involves identifying potential risks, evaluating their likelihood and impact, and developing strategies to mitigate them. This ensures that initiatives are not only feasible but also have a higher chance of successful implementation.

Risk assessment provides a structured approach to understanding the potential obstacles that could impede the success of digital transformation initiatives. By systematically evaluating risks, organizations can make informed decisions that balance innovation with caution. Here's a closer look at the steps involved in risk assessment:

Steps to Conduct a Risk Assessment

a. **Identify Potential Risks:** Recognize all possible risks that could affect the initiative, including technological, operational, financial, and market-related risks. For example, a CRM system implementation might face integration challenges with existing systems, high initial costs, employee resistance to change, and data privacy concerns.

b. **Evaluate Likelihood and Impact:** Assess the probability of each risk occurring and its potential impact on the initiative using a scoring system (e.g., 1 to 5). This helps prioritize risks that require immediate attention. For instance, integration challenges might score high in both likelihood and impact, making them a top priority for mitigation.

c. **Prioritize Risks:** Determine which risks require immediate action based on their scores. High-likelihood and high-impact risks need prompt mitigation strategies, while low-likelihood and low-impact risks require regular monitoring.

d. **Develop Mitigation Strategies:** Create strategies to reduce the likelihood or impact of high-priority risks. Technological risks might involve conducting thorough testing and using reliable vendors. For operational risks, providing comprehensive training and developing clear process documentation can be effective.

e. **Monitor and Review Risks:** Continuously monitor risks throughout the project lifecycle and adjust mitigation strategies as necessary. Regular risk review meetings and updates to the risk assessment matrix ensure ongoing management and responsiveness to new risks.

Risk assessment is essential in digital transformation to balance the excitement of innovation with the prudence of risk management. This approach not only safeguards the feasibility of initiatives but also enhances their chances of successful implementation. As a leader, embracing both the thrill of innovation and the discipline of risk assessment will position your organization for sustainable success in the dynamic digital landscape.

Resource Availability and Capacity: Ensuring Feasibility of Digital Transformation Initiatives

Resource availability and capacity assessment is a crucial step in digital transformation planning. This involves evaluating the financial, human, and technological resources required for each initiative and ensuring that these resources are sufficient and appropriately allocated. Without adequate resources, even the most promising initiatives can fail to deliver the desired outcomes.

Steps:

a. Assess current resource capacity (financial, human, technological).

> **Financial Resources**: Budget for technology acquisition, implementation, maintenance, and unforeseen expenses.

> **Human Resources**: Skilled personnel required to plan, execute, and support the initiative.

> **Technological Resources**: Hardware, software, infrastructure, and tools necessary for implementation.

b. Identify gaps in resources.

c. Prioritize initiatives that can be supported with available resources.

d. Develop a Resource Allocation Plan.

e. Monitor and Adjust Resource Allocation.

Stakeholder Impact and Buy-in: Gauging Support for Digital Transformation Initiatives

Stakeholder impact and buy-in are critical components of successful digital transformation initiatives. Understanding the needs, expectations, and influence of various stakeholders ensures that initiatives gain the necessary support and address key concerns, thereby enhancing the likelihood of successful implementation.

Steps:

a. Identify key stakeholders (executives, employees, customers, partners).

b. Engage stakeholders early to understand their needs and concerns.

c. Use stakeholder analysis to map and prioritize initiatives with broad support and positive impact.

d. Develop Engagement and Communication Strategies: Create tailored strategies to engage stakeholders, address their concerns, and gain their support.

e. Monitor and Foster Buy-in: Continuously monitor stakeholder engagement and foster ongoing buy-in throughout the initiative.

Regulatory and Compliance Requirements: Ensuring Adherence in Digital Transformation

Regulatory and compliance requirements are essential considerations in digital transformation initiatives. Adhering to these standards ensures legal compliance and helps maintain trust and credibility with stakeholders. Failure to comply can result in significant penalties, legal issues, and reputational damage.

In 2016, Uber concealed a data breach that exposed the personal information of 57 million users and drivers. Instead of disclosing the breach, Uber paid the hackers $100,000 to delete the data and keep the incident quiet (Uber Hid 2016 Breach, 2017). Consequently, Uber agreed to pay $148 million as part of a settlement with 50 U.S. states and Washington, D.C., for failing to disclose the breach (Kitty Paul, 2018)In the U.S. in 2018, it was revealed that Cambridge Analytica, a political consulting firm, had harvested personal data from millions of Facebook profiles without user consent. This data was used to influence political campaigns. The incident highlighted significant violations of data protection regulations, including GDPR. As a result, Facebook was fined $5 billion by the Federal Trade Commission (FTC) in the United States and faced a €500,000 fine by the UK's Information Commissioner's Office (ICO (BBC, Cambridge Analytica: Facebook row firm boss suspended, 2018)

Similarly, in the UK in 2018, British Airways suffered a data breach that affected around 380,000 transactions, exposing the personal and financial details of customers due to vulnerabilities in their website and mobile app. The breach violated GDPR requirements for data protection and security. The UK's Information Commissioner's Office (ICO) fined British Airways £20 million for failing to protect customers' data (BBC, British Airways fined £20m over data breach, 2020).While all these organizations were fined, the impact of these breaches extended beyond financial penalties. They also faced significant reputational damage and were forced to implement operational changes to address their security deficiencies.

Steps To Ensure Compliance:

  a. Identify mandatory regulatory and compliance requirements.

  b. Evaluate initiatives for compliance needs.

  c. Prioritize initiatives addressing critical regulatory requirements.

  d. Integrate Compliance into project planning.

  e. Monitor and maintain compliance.

Ensuring regulatory and compliance adherence is essential for the successful implementation of digital transformation initiatives. By systematically identifying applicable regulations, conducting thorough compliance assessments, developing robust strategies, integrating compliance into project planning, and maintaining continuous monitoring, organizations can mitigate legal risks and build trust with stakeholders.

As a leader, understanding and implementing this framework will enable you to guide your organization through a compliant and secure digital transformation journey, safeguarding against potential legal and reputational risks.

Return on Investment (ROI): Maximizing Financial Return from Initiatives

As a personal rule, I approach nearly every aspect of my life with a focus on ROI. Sometimes, my ROI is financial, while other times, it is intrinsic. As the saying goes, "What is worth doing is worth doing well."

Return on Investment (ROI) is a vital metric for evaluating the effectiveness of digital transformation initiatives. It measures the financial return generated by an investment relative to its cost. The main objective of maximizing ROI is to ensure that resources allocated to digital initiatives yield the highest possible financial benefits, thereby contributing to the organization's overall growth and profitability.

Steps to Maximize ROI:

a. Define clear objectives

b. Develop a financial model to estimate ROI for each initiative.

c. Identify Costs

d. Calculate ROI for both immediate and long-term benefits.

e. Formula:

$$ROI - \frac{\text{Net Benefits} - \text{Total Costs}}{\text{Total Costs}} \times 100$$

f. Prioritize initiatives with the highest ROI.

Maximizing ROI is essential for ensuring that digital transformation initiatives deliver significant financial benefits relative to their costs. By defining clear objectives and metrics, identifying all associated costs, estimating potential benefits, calculating ROI, conducting sensitivity analysis, and implementing and monitoring the initiatives, organizations can optimize their investments and achieve substantial returns.

As a leader, understanding and applying this framework will enable you to drive financially successful digital transformation initiatives that contribute to the long-term growth and profitability of your organization.

Customer Impact: Enhancing Customer Experience and Satisfaction

Customer impact is a crucial consideration in digital transformation initiatives. Enhancing customer experience and satisfaction can lead to increased loyalty, higher retention rates, and, ultimately, greater revenue. Understanding and addressing customer needs, preferences, and pain points is essential for delivering a superior customer experience.

Steps to enhance customer experience and satisfaction:

a. Understand Customer Needs and Expectations; collect customer feedback and data.

b. Map the Customer Journey and evaluate initiatives for their impact on customer experience.

c. Personalized Customer Interactions tailor interactions to individual customer preferences and behaviors.

d. Measure Customer Satisfaction.

e. Prioritize initiatives that deliver significant customer value.

Enhancing your customer experience and satisfaction is a vital objective of digital transformation initiatives. By understanding their needs and expectations, mapping the customer journey, personalizing interactions, improving customer support, and measuring satisfaction, organizations can create a superior customer experience that drives loyalty, retention, and revenue growth.

As a leader, focusing on customer impact will enable you to guide your organization toward delivering exceptional value and building lasting relationships with customers.

Time Sensitivity in Digital Transformation

Time sensitivity in digital transformation is critical for addressing urgent and timely initiatives. In today's fast-paced business environment, some projects cannot be delayed due to competitive pressures, regulatory deadlines, or rapidly evolving technologies. Prioritizing initiatives based on their strategic alignment, impact, and feasibility is essential.

For instance, **Walmart** recognized the urgent need to enhance its e-commerce capabilities to compete with Amazon. In response, Walmart accelerated its digital transformation efforts by acquiring Jet.com in 2016, a move that provided a significant boost to its online presence. They also implemented advanced data analytics to optimize inventory management and personalized customer experiences (Nanda, 2017). This quick and decisive action helped Walmart increase its online sales and improve customer satisfaction, demonstrating the importance of timely initiatives in digital transformation.

Another example is **Siemens**, a global industrial manufacturing company, which faced the urgent need to digitalize its operations to remain competitive in Industry 4.0. Siemens adopted MindSphere, an open IoT operating system, to connect their products, plants, systems, and machines, enabling them to harness the power of data. By integrating this technology quickly, Siemens was able to offer new digital services, improve operational efficiency, and open new revenue streams. This swift implementation underscored the critical role of time sensitivity in digital transformation to stay ahead in the market.

Steps to :

a. Identify time-sensitive initiatives with critical deadlines.

b. Assess the potential benefits of timely implementation.

c. Prioritize initiatives based on urgency and market opportunities.

These examples highlight the importance of addressing time-sensitive initiatives in digital transformation. By recognizing the urgency, prioritizing strategic initiatives, and leveraging technology, companies like Walmart and Siemens have successfully navigated their digital journeys, demonstrating the tangible benefits of timely and well-coordinated digital transformation efforts. Ultimately, addressing time sensitivity in digital transformation requires a strategic, agile, and well-coordinated approach to achieve timely and successful outcomes.

Competitive Advantage through Innovation

In today's fast-changing digital landscape, sustaining a competitive edge demands a proactive approach to innovation, ensuring the organization stays ahead of industry developments and meets evolving customer expectations. Innovation is essential for any dynamic organization, providing new avenues for growth and industry leadership.

It generally involves the introduction of new products and services to the market, making it a central focus of modern business research. Innovative ideas can greatly improve product quality and lower costs, potentially causing a significant shift in the market.

Innovation has the capacity to revolutionize traditional practices or even create entirely new and more effective markets. Without ongoing innovation, an organization risks falling behind in sales and profitability, making it susceptible to being overtaken by competitors. In essence, innovation is a fundamental driver of global competitiveness.

Steps for Fostering Innovation and Competitive Advantage:

a. Evaluate Innovation Potential: Identify initiatives with the potential to introduce new ideas, products, or processes that could significantly enhance the organization's offerings or disrupt the market.

b. Assess Unique Capabilities: Analyze how each initiative enhances the organization's unique capabilities or differentiates it from competitors, focusing on proprietary technologies, customer experience improvements, and scalability.

c. Prioritize Competitive Advantage: Prioritize initiatives that align with strategic objectives, offer significant market impact, can be quickly deployed, and provide a strong balance of risk and reward.

These steps ensure that the organization focuses on initiatives that drive innovation, enhance its unique strengths, and secure a sustainable competitive edge.

Implementing Filtering Strategies

Organizations often mistakenly view digital transformation as merely the integration of new technologies into their operations or the development and rollout of a technology product. However, effective digital transformation requires a more strategic approach that goes beyond technology deployment.

Earlier, we introduced and explored the fundamental concepts of filtering strategies. Implementing the filtering strategies outlined in this chapter can help organizations and transformation leaders prioritize the most impactful initiatives. To effectively put these strategies into action, organizations should adopt a structured, step-by-step approach that ensures alignment with strategic objectives and maximizes the impact of their digital transformation efforts.

- Define Criteria

Start by establishing clear, well-defined criteria that reflect the key components of a successful digital transformation. These criteria should be aligned with the organization's strategic goals and might include factors such as innovation potential, resource efficiency, risk level, and expected return on investment. By defining these criteria upfront, you create a solid foundation for evaluating and filtering potential initiatives, ensuring that resources are directed toward the most valuable opportunities.

- Gather Data:

With your criteria in place, the next step is to gather comprehensive data on each potential initiative. This data should include input from key stakeholders, financial projections, risk assessments, and a detailed analysis of resource requirements. The goal is to build a complete picture of each initiative, enabling a thorough and informed evaluation. Gathering accurate and relevant data is crucial, as it provides the basis for making objective decisions that align with the organization's strategic priorities.

- Evaluate and Prioritize:

Once the data is collected, use the defined criteria to systematically evaluate each initiative. This involves assessing the potential value of each initiative to the organization and ranking them accordingly. Prioritize initiatives that score high across multiple criteria, ensuring a balanced approach that considers both short-term gains and long-term strategic value. This step helps the organization allocate resources effectively, focusing on initiatives that are most likely to drive successful outcomes and deliver meaningful impact.

- Review and Adjust:

Regularly review and adjust the filtering criteria and the selected initiatives to maintain alignment with evolving organizational goals, market dynamics, and technological advancements. The business environment is constantly changing, so it's important to remain agile and responsive. This ongoing review process ensures that the organization can adapt to new opportunities and challenges as they arise, keeping the digital transformation journey on track and relevant.

- Communicate and Engage:

Finally, communication and engagement are critical to the successful implementation of these strategies. Clearly communicate the rationale behind the selected initiatives to all stakeholders, ensuring that everyone understands why certain initiatives were prioritized. Actively engage stakeholders throughout the process to foster support, collaboration, and shared ownership of the digital transformation journey. When stakeholders are informed and involved, they are more likely to support the initiatives and contribute to their success.

By following the Digital Transformation Filtering Framework, organizations can systematically prioritize high-impact initiatives, ensuring successful and sustainable transformation. This structured approach aligns with strategic goals, optimizes resource allocation, and supports continuous improvement. Regularly assessing cultural fit and organizational readiness ensures initiatives are well-integrated and embraced, leading to a unified approach to digital transformation.

Ultimately, embracing these filtering strategies simplifies complexity, enhances effectiveness, and keeps the organization competitive and innovative in the digital age.

Frameworks for Assessing Feasibility and ROI

Profitability is a key driver for most digital transformation initiatives. By focusing on projects that deliver a strong return on investment (ROI), organizations can ensure that their transformation efforts contribute positively to the bottom line.

To further enhance the effectiveness of the Digital Transformation Filtering Framework, organizations can implement additional frameworks for assessing the feasibility and return on investment (ROI) of their prioritized initiatives. These frameworks ensure that the selected initiatives not only align with strategic goals but are also practical and financially viable.

- Frameworks for Assessing Feasibility and ROI

1. Feasibility Assessment Framework:

- Technical Feasibility: Evaluate the technical requirements and capabilities needed to implement the initiative. Assess whether the existing infrastructure, technology stack, and expertise are sufficient or if additional investments are necessary. This step ensures that the initiative is technically achievable within the current organizational setup.

- Operational Feasibility: Consider the impact on day-to-day operations. Analyze whether the organization can integrate the initiative without disrupting core functions. Assess the readiness of operational processes and the capacity to manage change.

- Financial Feasibility: Review the financial implications, including upfront costs, ongoing expenses, and potential cost savings. Determine if the organization has the financial resources to support the initiative and whether it fits within the budgetary constraints.

- Legal and Regulatory Feasibility: Ensure that the initiative complies with relevant laws and regulations. Consider data protection, privacy, and industry-specific regulations that might affect implementation.

2. ROI Assessment Framework:

- Cost-Benefit Analysis: Perform a detailed cost-benefit analysis to compare the expected costs against the anticipated benefits of the initiative. This analysis should include both tangible (e.g., revenue growth, cost savings) and intangible benefits (e.g., customer satisfaction, brand value).

- Payback Period: Calculate the payback period, which is the time it will take for the initiative to generate enough benefits to cover its costs. This helps in understanding the short-term versus long-term financial impact.

- Net Present Value (NPV): Determine the NPV by discounting the future cash flows generated by the initiative to their present value. A positive NPV indicates that the initiative is expected to generate more value than it costs, making it a financially sound investment.

- Internal Rate of Return (IRR): Calculate the IRR, which represents the expected rate of return on the initiative. Compare this rate to the organization's required rate of return to assess whether the initiative meets financial expectations.

3. Risk Assessment Framework:

- Risk Identification: Identify potential risks associated with the initiative, including technical challenges, market uncertainties, and organizational resistance. Document these risks and categorize them by severity and likelihood.

- Risk Mitigation Strategies: Develop strategies to mitigate identified risks. This might include contingency plans, additional training, or phased implementation to reduce the impact of potential issues.

- Sensitivity Analysis: Conduct sensitivity analysis to understand how changes in key assumptions (e.g., costs, timelines, market conditions) might affect the outcome of the initiative. This helps prepare for different scenarios and make informed decisions.

Integration with the Digital Transformation Filtering Framework

By integrating these feasibility and ROI assessment frameworks into the Digital Transformation Filtering Framework, organizations can ensure that their prioritized initiatives are not only aligned with strategic goals but are also feasible, financially viable, and low risk. This comprehensive approach enhances decision-making, optimizes resource allocation, and increases the likelihood of successful and sustainable digital transformation outcomes.

This integrated process supports long-term success by focusing on initiatives that deliver significant value, are achievable within the organizational context, and align with financial expectations. Ultimately, this approach ensures that digital transformation efforts contribute to the organization's competitive advantage and growth in the digital era.

Part 3

Implementation and Execution

Chapter 5

Leadership and Change Management for Digital Transformation

Leadership plays a crucial role in driving digital transformation, a process that encompasses both people and technology - guiding the organization through the complexities of change.

Following the integration of profit considerations and organizational alignment, effective leadership and change management become crucial elements for the success of digital transformation initiatives. Leadership not only drives the vision and strategic direction but also plays a key role in managing the organizational change that digital transformation inevitably brings.

As a digital transformation leader, I've witnessed the crucial role that leadership plays in the success of these changes.

"Change Management: The People Side of Change" by Jeffrey M. Hiatt and Timothy J (Jeffrey M. Hiatt, 2003). Creasey focuses on the human aspects of organizational change. The book emphasizes that successful change isn't just about implementing new processes or technologies; it's about helping people transition from current ways of working to new ones. It introduces the ADKAR model (Awareness, Desire, Knowledge, Ability, Reinforcement) as a framework for understanding and managing individual change".

The authors highlighted the importance of leadership, communication, and addressing resistance, stressing that effective change management is essential for achieving sustainable transformation and realizing the full benefits of change initiatives.

In this chapter, I'll explore the essential role of leadership in driving digital transformation and managing change effectively.

Leading Digital Transformation

- ### *Articulating a Clear and Compelling Vision:*

A clear vision is the cornerstone of any successful digital transformation. This vision should not only define what the organization aims to achieve through digital transformation but also outline the path to get there. It needs to be both ambitious and realistic, inspiring the organization to move forward while also providing a clear direction. A compelling vision helps in aligning all stakeholders, from the C-suite to the operational teams, ensuring that everyone understands the goals and their role in achieving them.

Without a clear vision, digital transformation efforts can become fragmented, with different parts of the organization moving in different directions. A unified vision ensures that all digital initiatives are working towards a common goal, maximizing the impact of the transformation. Furthermore, a compelling vision helps to overcome resistance to change by painting a picture of a better future that everyone in the organization can buy into.

- ### *Aligning the Vision with Strategic Goals:*

Digital transformation should not be an isolated initiative; it must be deeply integrated with the organization's overall strategy. This means that the goals of digital transformation should directly support the broader business objectives, such as increasing market share, improving customer satisfaction, or enhancing operational efficiency. Leaders need to ensure that digital initiatives are not just about adopting new technologies but are aimed at driving strategic outcomes.

Aligning digital transformation with strategic goals ensures that the investments in technology are justified and contribute to the long-term success of the organization. This alignment also helps in securing buy-in from other executives and stakeholders, as they can clearly see how digital initiatives will support the overall business strategy.

- ***Communicating the Vision and Long-Term Benefits:***

Once the vision is articulated, it's crucial to communicate it effectively across the organization. Leaders must ensure that the vision is understood at all levels, from top management to frontline employees. This communication should highlight not just the immediate benefits but also the long-term advantages of digital transformation, such as staying competitive, enhancing customer experiences, and driving innovation.

Effective communication helps to build momentum and keep the organization focused on the transformation goals. It also helps manage expectations and reduce uncertainty, which can be significant barriers to change. By clearly communicating the long-term benefits, leaders can motivate employees to embrace the transformation, even if it involves short-term disruptions.

- ***Empowering Teams through Delegation:***

The idea that "if you want something done right, you have to do it yourself" is a mindset that doesn't align with successful digital transformation. Digital transformation cannot be achieved by leadership alone; it requires active participation from teams across the organization. Leaders must delegate authority and responsibility to those who are directly engaged with the work, empowering them to make decisions and lead initiatives. This empowerment must be supported with the necessary resources, training, and guidance to ensure that teams are equipped to execute effectively.

Empowered teams are more agile and can respond more quickly to changes and challenges. This agility is crucial in a fast-paced digital environment where quick decisions can make the difference between success and failure. Empowerment also fosters a sense of ownership and accountability, encouraging teams to take initiative and innovate.

-   Fostering a Culture of Creativity and Experimentation

Achieving successful digital transformation often requires organizations to adopt new ways of thinking and experimenting with different approaches to work. Leaders should foster an environment where creativity is encouraged, and experimentation is supported, even when it involves taking calculated risks. This means giving teams the freedom to explore new ideas, learn from their mistakes, and quickly iterate on their efforts.

In one of the organizations where I worked, digital transformation thrived by promoting creativity through initiatives like hackathons. A culture that embraces creativity and experimentation is essential for driving innovation, which is central to digital transformation. Organizations that encourage this kind of experimentation are more likely to uncover new opportunities and solutions that propel their transformation efforts. Additionally, this culture helps to overcome the fear of failure, which can otherwise hinder innovation and slow down the transformation process.

Managing Organizational Change in Digital Transformation

-   Addressing the Human Side of Change:

As I noted in the introduction, digital transformation involves people just as much as it involves technology. For a transformation to be successful, it's crucial to address the concerns and needs of employees who are impacted by the change. This means offering clear communication, training, and support to help them shift from their current ways of working to new approaches. Leaders must also be empathetic, recognizing that change can be challenging for many individuals.

Ignoring the human side of change can lead to resistance, low morale, and even failure of the transformation efforts. By focusing on the people aspect, leaders can ensure a smoother transition with employees who are more engaged and willing to support the transformation.

- *Applying the ADKAR Model:*

The ADKAR model (Awareness, Desire, Knowledge, Ability, Reinforcement) is a framework for managing individual change. It provides a structured approach to help employees move through the stages of change. Leaders can use this model to identify where individuals are in the change process and what support they need to move forward.

The ADKAR model helps in breaking down the change process into manageable steps, making it easier to address individual concerns and ensure that employees are fully on board with the transformation. This approach also helps identify potential roadblocks early and address them before they become major issues.

- *Leadership in Crisis Management:*

Digital transformation often comes with its share of crises, whether it's a failed system implementation, cybersecurity threats, or unexpected resistance to change. Leaders need to be prepared to handle these crises effectively, using them as opportunities to reinforce the importance of the transformation and demonstrate resilience.

How leaders handle crises during digital transformation can significantly impact the success of the initiative. Effective crisis management can turn a potential setback into a learning opportunity, strengthening the organization's resolve to continue with the transformation.

- *Building Relationships and Collaboration:*

Digital transformation requires collaboration across different functions and departments. Leaders must build strong relationships and foster a collaborative culture where everyone is working towards a common goal. This includes breaking down silos, encouraging cross-functional teams, and ensuring that all parts of the organization are aligned with the transformation objectives.

Collaboration is essential for overcoming the challenges of digital transformation. When teams work together, they can share knowledge, solve problems more effectively, and create innovative solutions that drive the transformation forward.

- *Continuous Learning and Adaptability:*

Digital transformation is a continuous journey that demands ongoing learning and adaptation. Leaders should promote a culture of continuous improvement, where employees are encouraged to learn new skills, stay updated with the latest technologies, and adapt to changing circumstances. This also involves leaders themselves being open to learning and adapting their strategies as needed.

In a rapidly changing digital landscape, the ability to learn and adapt quickly is crucial for sustaining the momentum of transformation. Continuous learning ensures that the organization remains competitive and can take advantage of new opportunities as they arise.

Leadership is at the heart of successful digital transformation. It involves more than just setting the direction; it requires guiding the organization through the complexities of change, empowering teams, fostering a culture of innovation, and managing the human aspects of transformation. By focusing on these expanded leadership principles, transformation leaders can navigate the challenges of digital transformation and achieve sustainable success.

Strategies for Fostering a Culture of Innovation and Managing Change

How can you cultivate a culture of innovation within your organization and among your team?

- *Encourage a Growth Mindset:*

A culture of innovation begins with cultivating a growth mindset throughout the organization. Leaders should encourage employees to view challenges and failures as opportunities to learn and grow rather than as setbacks. This mindset shift is crucial because innovation often involves taking risks and experimenting with new ideas, which can sometimes lead to failure.

By fostering a growth mindset, leaders can create an environment where employees feel safe to take risks, experiment, and explore new ideas without fear of negative consequences. This, in turn, drives creativity and leads to breakthrough innovations.

- ***Create Cross-Functional Teams:***

Innovation thrives when diverse perspectives come together to solve problems. Establishing cross-functional teams that bring together individuals from different departments and disciplines can spark creative thinking and lead to innovative solutions. These teams should be empowered to collaborate, share knowledge, and challenge each other's ideas constructively. By breaking down silos and promoting collaboration across the organization, leaders can harness the collective intelligence of the workforce and drive innovation at all levels.

Additionally, cross-functional teams can help ensure that new ideas are practical and aligned with the organization's strategic goals, as they bring a holistic view to problem-solving.

- ***Provide Resources and Support for Experimentation:***

For innovation to flourish, employees need access to the resources, tools, and support necessary to experiment with new ideas. This includes not only financial resources but also time, technology, and mentorship. Leaders should establish a framework for experimentation that allows teams to pilot new ideas on a small scale before committing to full-scale implementation.

This approach reduces the risk associated with innovation while providing valuable insights into what works and what doesn't. By supporting experimentation, leaders demonstrate their commitment to innovation and encourage employees to think creatively and push boundaries.

- ***Recognize and Reward Innovation:***

Recognition and rewards play a vital role in fostering a culture of innovation. Leaders should actively recognize and celebrate employees who contribute innovative ideas or who take initiative in driving change. This recognition can take various forms, such as public acknowledgment, financial incentives, or opportunities for career advancement.

By highlighting and rewarding innovative behavior, leaders reinforce the importance of innovation within the organization and motivate others to contribute their ideas. Moreover, recognition helps to build a sense of pride and ownership among employees, further embedding innovation into the organizational culture.

- ### *Communicate a Clear Vision for Change:*

Managing change effectively requires leaders to communicate a clear and compelling vision for the future. This vision should articulate the purpose and benefits of the change, making it easy for employees to understand why the transformation is necessary and how it will impact them. Effective communication involves not only sharing the vision but also actively listening to employees' concerns and feedback.

Leaders should engage with employees at all levels, providing them with the information they need to embrace the change and aligning the vision with their personal and professional goals. This approach helps to build trust and ensures that employees are on board with the transformation.

- ### *Address Resistance with Empathy and Support:*

Resistance to change is a natural human response, and leaders must address it with empathy and support. Instead of dismissing concerns or pushing forward without regard to employee sentiments, leaders should seek to understand the root causes of resistance.

This may involve engaging in open dialogues with employees, providing additional training or resources, or adjusting the change plan based on employee feedback. By addressing resistance proactively and empathetically, leaders can reduce friction, build stronger relationships with employees, and increase the likelihood of successful change adoption.

- ### *Build Change Management Capabilities:*

Effective change management requires specific skills and capabilities, both at the leadership level and throughout the organization. Leaders should invest in developing these capabilities by providing training and resources on change management best practices, such as the ADKAR model or other frameworks.

Additionally, leaders should establish a dedicated change management team or task force responsible for guiding the organization through the transformation process. By building change management capabilities, leaders ensure that the organization is well-equipped to handle the complexities of change and can navigate the transformation smoothly.

- *Foster Continuous Learning and Adaptability:*

Innovation and change are not one-time events; they are ongoing processes that require continuous learning and adaptability. Leaders should encourage a culture of continuous improvement, where employees are motivated to keep learning, stay updated on industry trends, and adapt to new technologies and methods. This can be achieved through regular training sessions, access to online learning platforms, and creating opportunities for employees to experiment with new ideas.

By fostering continuous learning, leaders help the organization remain agile and responsive to emerging challenges and opportunities, ensuring sustained success in an ever-evolving digital landscape.

Chapter 6

Data-Driven Decision Making

In today's digital age, data is often referred to as the new oil—a critical asset that fuels decision-making, drives innovation, and creates a competitive edge. However, just as raw oil needs refining, data requires proper management and governance to be useful.

This chapter explores the pivotal role of data analytics in shaping business strategies and underscores the importance of data governance and quality management in ensuring that data serves as a reliable foundation for decision-making.

The Role of Data Analytics in Informing Strategies

- *Data-Driven Decision Making:*

Data analytics has revolutionized how organizations make decisions. By analyzing large volumes of data, businesses can uncover trends, patterns, and insights that were previously hidden or inaccessible. These insights enable leaders to make informed decisions that are based on evidence rather than intuition or guesswork.

For instance, Amazon uses advanced data analytics to personalize the shopping experience for each customer. By analyzing browsing history, past purchases, and search queries, Amazon's recommendation engine suggests products tailored to individual preferences, driving customer engagement and increasing sales.

Moreover, data analytics supports predictive and prescriptive analytics, allowing organizations to anticipate future trends and prescribe actions that will lead to desired outcomes. Netflix is a prime example of this, as it uses data analytics to make decisions about content creation. By analyzing viewer data such as watch history and interaction patterns, Netflix decides which content to produce, resulting in successful original content that resonates with audiences. This capability is particularly valuable in dynamic industries where the ability to foresee market shifts can provide a significant competitive advantage.

- *Enhancing Operational Efficiency:*

Data analytics plays a critical role in enhancing operational efficiency by identifying inefficiencies and optimizing processes.

For example, Walmart uses big data analytics to optimize its inventory management. By analyzing customer purchasing patterns, weather data, and social media trends, Walmart can predict demand for products in specific stores, ensuring that popular items are always in stock while minimizing overstock of less popular items. This data-driven approach reduces costs and improves overall efficiency and customer satisfaction.

In addition, data analytics enables real-time monitoring and performance tracking, allowing organizations to respond quickly to any issues that arise. This agility is essential in today's fast-paced business environment, where delays or inefficiencies can lead to lost opportunities or customer dissatisfaction.

- *Driving Innovation and Competitive Advantage:*

Innovation is at the heart of business growth, and data analytics is a key driver of innovation. By analyzing market trends, consumer behavior, and technological advancements, organizations can identify new opportunities for innovation. Spotify exemplifies this by using data analytics to power its music recommendation engine, helping users discover new music through personalized playlists like *Discover Weekly*. This data-driven personalization has significantly increased user engagement, keeping Spotify at the forefront of the streaming industry.

Data analytics also provides a competitive advantage by enabling businesses to differentiate themselves from competitors. Companies that effectively leverage data to inform their strategies are better positioned to anticipate market changes, meet customer needs, and respond to competitive threats.

- ***Strategic Risk Management:***

In addition to driving growth and efficiency, data analytics is crucial for managing risk. By analyzing historical data and identifying patterns, organizations can predict potential risks and develop strategies to mitigate them. For example, in the financial industry, data analytics is used to detect fraudulent activities or assess credit risk. In manufacturing, it helps predict equipment failures and prevent costly downtime.

Tesla is another great example; it uses data collected from its vehicles to improve its autonomous driving capabilities. By analyzing real-world driving data, Tesla can continuously refine its algorithms, enhancing the safety and reliability of its vehicles and positioning itself as a leader in autonomous driving technology.

Furthermore, data analytics supports scenario analysis, which allows organizations to explore different outcomes based on various assumptions. This capability enables businesses to prepare for a range of possible future scenarios and make strategic decisions that minimize risk.

Data Governance and Quality Management as a Catalyst for Digital Transformation

High-quality data is the foundation of effective data analytics. Without accurate, complete, and consistent data, the insights derived from analytics can be misleading, leading to poor decision-making. Data governance involves establishing policies, procedures, and standards to ensure that data quality is maintained throughout its lifecycle. This includes data collection, storage, processing, and analysis.

Effective data governance requires organizations to define data ownership, establish data quality metrics, and implement regular data audits to identify and address quality issues. By ensuring that data is reliable and trustworthy, organizations can be confident that their analytics are based on solid foundations.

In an era where data breaches and privacy concerns are increasingly common, data governance plays a critical role in protecting sensitive information. Data governance frameworks should include robust security measures to prevent unauthorized access, data breaches, and data loss. This involves implementing encryption, access controls, and regular security assessments.

Additionally, data governance ensures compliance with data protection regulations, such as the General Data Protection Regulation (GDPR) in Europe or the California Consumer Privacy Act (CCPA) in the United States. These regulations require organizations to handle personal data responsibly and provide individuals with certain rights regarding their data. Effective data governance helps organizations avoid legal penalties and maintain customer trust.

Data governance is not only about protecting data but also about ensuring that it is properly managed and accessible to those who need it. This includes establishing data management practices that support data integration, data sharing, and data reuse across the organization. By breaking down data silos and ensuring that data is easily accessible, organizations can maximize the value of their data assets.

It also involves creating a data catalog that provides metadata and context about the data, making it easier for users to find and understand the data they need. This is particularly important in large organizations with complex data environments, where data is often spread across multiple systems and formats.

For data analytics to be truly effective, organizations must foster a data-driven culture where data is valued and used as a key resource in decision-making. Data governance supports this culture by establishing clear guidelines for data usage, ensuring that employees have access to high-quality data, and providing training on data literacy.

A strong data governance framework also promotes accountability and transparency in data management, which helps build trust in data-driven decisions. When employees trust the data they are using, they are more likely to embrace data-driven practices and contribute to the organization's data-driven culture.

The Intersection of Data Analytics and Data Governance

The effectiveness of data analytics is directly linked to the strength of an organization's data governance practices. Without proper governance, data analytics initiatives can suffer from poor data quality, security risks, and compliance issues. Therefore, it is essential to align data analytics with data governance to ensure that analytics efforts are built on a solid foundation.

Organizations should establish clear processes for integrating data governance into their analytics workflows. This includes setting up data governance committees, defining data stewardship roles, and ensuring that data governance policies are consistently applied throughout the analytics lifecycle.

Data governance and analytics should not be static processes; they require continuous improvement. Organizations should implement feedback loops that allow them to monitor the effectiveness of their data governance practices and adjust as needed. This could involve regular reviews of data quality metrics, security assessments, and compliance checks.

Additionally, insights gained from data analytics can be used to inform and refine data governance practices. For example, if analytics reveals issues with data quality or accessibility, these insights can be used to update data governance policies and improve data management processes.

Data analytics is a powerful tool that enables organizations to make informed decisions, enhance operational efficiency, drive innovation, and manage risk. However, the effectiveness of data analytics depends on the quality, security, and management of the underlying data. This is where data governance and quality management come into play. By implementing robust data governance frameworks and maintaining high data quality standards, organizations can ensure that their analytics efforts are successful and that they can fully leverage the value of their data.

In today's data-driven world, the intersection of data analytics and data governance is critical to achieving strategic objectives and maintaining a competitive edge. As Leaders, you must prioritize both aspects to build a strong foundation for your organization's digital transformation journey.

Part 4

Advancing Through Collaboration and Future Outlook

Chapter 7

Building Partnerships and Ecosystems

No organization can thrive in isolation. Success in digital transformation increasingly depends on the ability to build strong partnerships and ecosystems that enable collaboration, innovation, and agility.

In this chapter, we will explore the importance of collaborative approaches to digital transformation and provide strategies for managing technology vendors and ecosystems. By leveraging these partnerships effectively, organizations can accelerate their transformation journeys and achieve sustainable competitive advantages.

- *The Power of Collaboration*

Digital transformation is a multifaceted process that often demands expertise and resources that extend beyond what any one organization can provide on its own. Collaborating with external partners allows organizations to access a broader spectrum of knowledge, technology, and innovation.

A prime example of this is Apple's App Store, which demonstrates how collaboration can fuel digital success. By opening its platform to third-party developers, Apple established an ecosystem that fosters innovation, resulting in the creation of millions of apps that, in turn, increase the value of its devices.

- *Strategic Alliances for Innovation*

Strategic alliances between companies can be a powerful driver of innovation in digital transformation. These partnerships allow organizations to combine their strengths and capabilities to develop new products, services, or business models. A notable example is the partnership between IBM and Maersk to develop a blockchain-based platform for global trade (Why The Commanders' Owner Thinks $6 Billion For An NFL Team Is A Bargain, 2024). This collaboration brought together IBM's expertise in blockchain technology and Maersk's extensive knowledge of the shipping industry to create a solution that improves transparency and efficiency in supply chains.

Similarly, the partnership between Microsoft and Adobe is another example of a strategic alliance that fosters innovation. By integrating Adobe's marketing tools with Microsoft's cloud platform, the two companies provide a comprehensive solution for businesses to manage their customer relationships more effectively, thereby driving digital transformation in marketing and customer engagement.

- ### *Co-Innovation with Customers*

Collaborating with customers can also be a key component of digital transformation. Co-innovation involves working closely with customers to develop new solutions that meet their specific needs. This approach not only ensures that the products or services are tailored to market demands but also strengthens customer relationships.

GE's Digital Foundry is an example of this approach, where the company collaborates with industrial customers to develop customized digital solutions that improve operational efficiency and performance.

- ### *Public-Private Partnerships*

In some cases, digital transformation initiatives require collaboration between public and private entities. Public-private partnerships (PPPs) can be particularly effective in sectors like healthcare, transportation, and smart cities, where the integration of technology and infrastructure is crucial.

For example, Highway 407 is a privately operated toll highway in Ontario. The government of Ontario entered into a PPP agreement with a private consortium to design, build, and operate the highway. The highway was one of the first major PPP projects in Canada. The Highway has been successful in providing a reliable alternative route for drivers in the Greater Toronto Area. The PPP model allowed for the efficient construction and operation of the highway while generating revenue through tolls.

Managing Technology Vendors and Ecosystems

Selecting the right technology vendors is crucial for the success of digital transformation initiatives. Organizations should perform comprehensive due diligence to evaluate the capabilities, reliability, and track record of potential vendors. It's essential to consider not only the technical features of the vendor's solutions but also their ability to align with the organization's long-term strategic goals.

For instance, in its digital transformation journey, General Electric (GE) chose Amazon Web Services (AWS) as its preferred cloud provider. GE opted for AWS due to its scalability, global presence, and robust support for industrial applications, which aligned well with GE's strategic objectives (Fitchard, 2017).

Once the right vendors are selected, building strong, collaborative relationships with them is essential. This involves establishing clear communication channels, setting mutual expectations, and fostering a partnership mentality rather than a transactional one.

How can you effectively oversee Ecosystem Management and Strategic Alignment in Digital Transformation?

Managing a technology ecosystem requires careful coordination of multiple vendors and partners to ensure that all components work harmoniously together. This holistic approach to integration, interoperability, and governance is essential for the smooth operation of complex digital environments. One of the most successful examples of effective ecosystem management is the Android operating system. Google oversees a vast ecosystem that includes device manufacturers, app developers, and service providers, ensuring that the Android platform remains consistent and compatible across a wide range of devices and regions. This comprehensive management strategy has been instrumental in making Android the most widely used mobile operating system globally.

In addition to managing ecosystems, organizations are increasingly leveraging open innovation platforms to accelerate their digital transformation efforts. These platforms facilitate collaboration with external developers, startups, and other partners, enabling the co-creation of new solutions.

Ensuring that technology vendors are aligned with an organization's strategic goals is another critical aspect of successful digital transformation. This alignment should be evident in the vendor's commitment to supporting the organization's objectives, as well as in the contractual agreements and performance metrics. For instance, when Starbucks embarked on its digital transformation journey, it strategically partnered with Microsoft to leverage its cloud and AI capabilities. The synergy between Starbucks' goal of enhancing customer experience and Microsoft's technological expertise played a pivotal role in the success of digital initiatives such as personalized customer recommendations and the mobile ordering app.

Risk management is also a crucial component of maintaining healthy vendor relationships within a technology ecosystem. Organizations must regularly assess their vendors' performance and security measures to mitigate potential risks. This includes conducting audits, monitoring service levels, and having contingency plans in place. For example, financial institutions often engage multiple technology vendors to ensure redundancy and minimize the risk of service disruptions. By diversifying their vendor base and maintaining strict oversight, these institutions protect themselves against potential risks while continuing to innovate and adapt to changing market conditions.

Building strong partnerships and managing technology ecosystems effectively are essential components of successful digital transformation. Collaborative approaches, whether through strategic alliances, customer co-innovation, or public-private partnerships, enable organizations to access new capabilities, drive innovation, and achieve their transformation goals.

At the same time, managing technology vendors and ecosystems requires careful selection, strong relationship-building, and a focus on alignment with strategic objectives.

In today's interconnected digital world, no organization can afford to go it alone. By leveraging partnerships and ecosystems, organizations can accelerate their digital transformation journeys and position themselves for long-term success in an increasingly competitive landscape.

Chapter 8

Measuring Success and Iterating

If success cannot be measured, it becomes a matter of personal perception. Measuring success and iterating are vital elements of an effective digital transformation. Without clear measurements, success remains a vague concept, open to subjective interpretation, which can lead to confusion and misalignment of goals.

The first step is to define what success means by setting clear metrics that align with your organization's strategic goals. Metrics serve as tools to track progress, demonstrate value, and identify areas for improvement. These can include key performance indicators (KPIs) like cost savings, productivity gains, customer satisfaction, or quicker time-to-market for new products.

Moreover, focusing on specific business outcomes such as increased revenue, enhanced operational efficiency, or improved customer experiences helps anchor transformation efforts in concrete results. Employee engagement and customer-related metrics are also crucial in assessing the overall impact of digital initiatives.

Ongoing measurement is critical to understanding progress and making necessary adjustments. Real-time dashboards can be used to monitor current initiatives, providing visibility into key metrics and enabling rapid responses to new challenges. Establishing feedback loops with stakeholders such as employees, customers, and partners helps gather valuable insights on what's working and where improvements are needed. Benchmarking against industry standards or historical data can further contextualize the success of your digital transformation efforts.

Iteration is where the real transformation happens. Adopting an agile methodology allows organizations to break down large initiatives into smaller, manageable projects, enabling faster iterations and adjustments based on feedback and data. Pilot programs are particularly useful in this regard; they allow you to test new digital initiatives, measure their success, and iterate before scaling them up. Learning from failures is equally important, as it provides opportunities to refine strategies and try again with a more informed approach. This mindset of continuous, incremental improvement ensures that processes, technologies, and strategies are constantly being optimized based on ongoing measurement and insights.

Sustaining digital transformation requires keeping a long-term vision in mind, where success is not only measured by immediate gains but also by the sustainability and scalability of initiatives. Cultural shifts within the organization, such as fostering innovation or moving towards data-driven decision-making, should be measured to assess the broader impact of the transformation. For example, internal hackathons can be a powerful way to measure creativity, engagement, and innovation, tracking how many ideas are successfully implemented and their subsequent impact on the business.

Measuring success and iterating in digital transformation is a dynamic process that requires a combination of well-defined metrics and a willingness to adapt based on data-driven insights. This continuous cycle of evaluation and refinement ensures that digital initiatives not only meet immediate goals but also contribute to long-term organizational growth and resilience.

Establishing KPIs and Metrics for Measuring Success

Establishing Key Performance Indicators (KPIs) and metrics is crucial for measuring success and ensuring that your efforts are driving meaningful outcomes. KPIs and metrics provide a structured way to evaluate progress, make informed decisions, and demonstrate value to stakeholders. Here's how to effectively establish and utilize these tools in your digital transformation journey.

- *Align with Strategic Objectives*

The first step in establishing KPIs and metrics is to ensure they are closely aligned with your organization's strategic objectives. Consider what your organization aims to achieve through digital transformation, whether it's enhancing customer experience, improving operational efficiency, driving innovation, or increasing revenue. Your KPIs should directly reflect these goals, providing a clear line of sight between your transformation initiatives and the broader business strategy.

- *Define Specific and Measurable KPIs*

KPIs should be specific, measurable, attainable, relevant, and time-bound (SMART). For example, if one of your objectives is to improve customer satisfaction, a specific KPI might be to increase the Net Promoter Score (NPS) by a certain percentage within a defined timeframe. Other examples of specific KPIs might include reducing operational costs by a certain amount, decreasing the time-to-market for new products, or increasing the adoption rate of digital tools among employees.

- *Consider a Balanced Approach*

It's important to take a balanced approach when selecting KPIs. Include a mix of financial, operational, and customer-centric metrics to gain a comprehensive view of your digital transformation's impact. Financial KPIs might track revenue growth, cost savings, or return on investment (ROI). Operational KPIs could measure process efficiency, employee productivity, or system uptime. Customer-centric KPIs might include customer satisfaction scores, digital engagement levels, or retention rates.

- *Leading and Lagging Indicators*

In addition to traditional lagging indicators that measure outcomes after they've occurred (e.g., revenue growth), consider incorporating leading indicators that predict future performance. Leading indicators might include the rate of digital adoption, the number of new ideas generated through innovation programs, or the speed of response to customer inquiries. These indicators can provide early signals of success or areas needing improvement, allowing for proactive adjustments.

- *Implement Real-Time Monitoring*

Real-time monitoring tools and dashboards are essential for continuously tracking KPIs and metrics. These tools provide up-to-date insights into your digital transformation efforts, enabling quick decision-making and the ability to respond to emerging challenges or opportunities. Real-time data helps ensure that your organization remains agile and capable of adjusting strategies as needed.

- *Involve Stakeholders in KPI Development*

Engage key stakeholders such as department heads, team leaders, and even customers in the development of KPIs and metrics. This ensures that the chosen KPIs are relevant to all areas of the organization and that there is buy-in across teams. Involving stakeholders also helps ensure that KPIs reflect the perspectives and needs of those who are directly impacted by the digital transformation.

- *Regularly Review and Adjust KPIs*

Digital transformation is an ongoing process, and your KPIs and metrics should evolve as your organization grows and adapts. Regularly review your KPIs to ensure they remain aligned with your strategic objectives and are still relevant in the context of your transformation efforts. As new technologies, market conditions, or business priorities emerge, be prepared to adjust your KPIs accordingly.

- **Communicate Success and Insights**

Finally, use the data gathered from your KPIs and metrics to communicate success and insights to stakeholders. This not only demonstrates the value of your digital transformation efforts but also fosters a culture of transparency and continuous improvement. Regular reporting and sharing of KPI results help keep everyone aligned and motivated toward achieving the organization's transformation goals.

Establishing and utilizing KPIs and metrics is a critical component of measuring success in digital transformation. By aligning these indicators with strategic objectives, ensuring they are specific and measurable, and regularly reviewing and adjusting them, you can effectively track progress, drive informed decision-making, and achieve lasting success in your digital transformation journey.

Continuous Improvement Towards Long-Time Success

What methodology is your organization using, and how successful is it in your transformation journey?

Continuous improvement is a crucial philosophy for organizations that aim to sustain long-term success, particularly in dynamic and competitive environments. This approach focuses on incremental enhancements to your processes, products, or services, fostering innovation, efficiency, and adaptability.

Several methodologies have been developed to guide organizations in their continuous improvement efforts. Let's take a look at some of the most widely used methodologies that every digital transformation leader should be familiar with and apply to reduce efficiencies. The method you choose for your organization depends on your overall transformation goals, as discussed in Chapter 3 - The Maze of Choices: Technologies and Strategies.

- *Six Sigma*

General Electric (GE) famously adopted Six Sigma under the leadership of former CEO Jack Welch. GE used Six Sigma to drive quality improvements and cost reductions across the organization, which resulted in billions of dollars in savings. The focus on reducing defects and optimizing processes significantly enhanced customer satisfaction and operational efficiency.

Six Sigma is a data-driven methodology aimed at reducing defects and variability in processes. It uses a structured approach known as DMAIC (Define, Measure, Analyze, Improve, Control) to identify and eliminate the root causes of problems.

By focusing on quality improvement and consistency, Six Sigma helps organizations like GE and Motorola achieve higher levels of customer satisfaction and operational efficiency. This methodology relies on statistical tools and techniques to analyze data, which informs decisions about process improvements.

Table 8.1: Definition of DMAIC

D – Define the opportunity
M – Measure the performance of existing processes
A – Analyze the process to find defects and root causes
I – Improve the process by addressing root causes.
C – Control any improved process and assess future process performance to correct deviations.

Six Sigma is frequently applied in manufacturing because it focuses on reducing defects and variability. The primary objective is to achieve consistency, which ultimately enhances customer satisfaction.

- *Total Quality Management (TQM)*

Total Quality Management is a customer–focused method that involves continuous improvement and seeks to improve the quality of all organizational processes, with a focus on long-term success through customer satisfaction.

Sony (TQM's Challenge to Management Theory and Practice, 1994) applied TQM practices to improve product quality and operational efficiency across its global operations. By focusing on customer satisfaction and continuous improvement, Sony has been able to maintain its reputation for innovation and high-quality products.

TQM involves every employee in the organization, from top management to frontline workers, in a collective effort to improve processes, products, and services. This methodology emphasizes continuous assessment and improvement, using tools like process mapping, benchmarking, and root cause analysis to drive quality enhancements.

- *Learn*

Lean methodology focuses on maximizing value for the customer while minimizing waste. Originating from the Toyota Production System, Lean emphasizes the elimination of non-value-added activities, streamlining processes, and enhancing efficiency. Key principles of Lean include defining value from the customer's perspective, mapping the value stream to identify and remove waste, creating flow in processes, and seeking perfection through continuous refinement.

Lean is widely used in manufacturing but has also been successfully applied in various industries, including healthcare, finance, and services.

- *Kaizen &Continuous Improvement Process (CIP)*

Nestlé applies both Kaizen and CIP methodologies across its global operations to improve efficiency and reduce waste (Nestlé reports full-year results for 2020, 2021). Kaizen, a Japanese term meaning "change for the better," and the Continuous Improvement Process (CIP) are closely related philosophies in continuous improvement. Both emphasize the importance of making small, incremental changes that, over time, lead to significant enhancements in efficiency, quality, and overall performance.

The company encourages employees to identify areas for improvement in their daily work processes, which has led to significant enhancements in productivity and product quality. Nestlé's commitment to continuous improvement is a key factor in its ability to remain competitive in the global market.

This methodical approach ensures that even minor adjustments contribute to the organization's long-term success by reducing waste and enhancing quality.

- *PDCA (Plan-Do-Check-Act)*

PDCA, also known as the Deming Cycle, is a straightforward and iterative four-step process for continuous improvement.

Table 8.2: The Four Processes of PDCA:

P – Plan phase, where goals and processes are defined based on data and analysis
D - In the do phase, the plan is implemented on a small scale to test its effectiveness.
C - The check phase involves evaluating the outcomes against the expected results and identifying any discrepancies.
A - The act phase involves making necessary adjustments based on the evaluation and standardizing successful practices.

Nike has used the PDCA cycle in its sustainability initiatives. For instance, when developing their sustainable product lines, Nike planned by setting goals for reducing the environmental impact of their products, executed changes by incorporating recycled materials into their production, checked the effectiveness of these changes through rigorous testing and feedback and acted by scaling up successful strategies across their product lines.

- *Agile*

Although Agile methodology was originally developed for software development, it has since become a widely adopted approach for continuous improvement across various industries. Over the past decade, I have extensively utilized this methodology in the transformations I've led, successfully applying it to drive significant change.

Agile methodology prioritizes flexibility, customer collaboration, and iterative progress. In an Agile framework, projects are divided into small, manageable increments known as sprints. After each sprint, teams review the outcomes, gather feedback, and make necessary adjustments before moving forward. This iterative process ensures that improvements are continuously integrated into the project, allowing for rapid adaptation to changing conditions and customer needs.

Organizations like Spotify, ING Bank, Wexner Medical Center, and John Deere have successfully applied Agile, demonstrating its versatility across different sectors. These examples highlight how Agile can be adapted to fit specific organizational needs, resulting in faster innovation, improved customer satisfaction, and better alignment with strategic goals.

- *Business Process Reengineering (BPR)*

As a digital transformation leader, when I join an organization, I often find that enhancing existing business processes is a quick win before embarking on any major transformation initiatives. This approach typically leads to immediate efficiency gains, as it allows for the rapid optimization of current operations before implementing broader changes.

Business Process Reengineering focuses on fundamentally rethinking and redesigning business processes to achieve significant improvements in critical performance measures, such as cost, quality, and speed.

Unlike other continuous improvement methodologies that emphasize incremental change, BPR advocates for radical change to achieve dramatic improvements. It involves a thorough analysis of existing processes, identifying bottlenecks and inefficiencies, and redesigning processes from the ground up to better meet organizational goals.

Overall, continuous improvement methodologies offer structured approaches that enable organizations to progressively enhance their processes, products, and services. Whether through the incremental adjustments characteristic of Kaizen and Lean or the more radical restructuring found in Business Process Reengineering (BPR), these methodologies support organizations in staying competitive, agile, and responsive to evolving market demands.

By embedding continuous improvement into the organizational culture, transformation leaders can drive sustained success and foster ongoing innovation within their companies.

Chapter 9

Emerging Trends and Future Considerations

In Chapter 3, we examined a range of technologies that are transforming the modern business landscape. Now, in this chapter, we will take a closer look at the key technologies driving digital transformation, focusing on their effects and practical uses. We will explore how these innovations are not only disrupting established business models but also opening up new opportunities and presenting unique challenges.

While many organizations acknowledge the importance of these technologies, their practical implementation remains a challenge. This chapter not only outlines the potential of these innovations but also provides an exploration of how they can enhance existing business processes, enabling companies to thrive in a digital-first landscape.

By understanding their roles and implementations, we aim to provide a comprehensive view of how these innovations are transforming industries and influencing strategic decisions.

Unpacking Emerging Disruptive Technologies and Their Impact on Digital Transformation

1. Artificial Intelligence (AI) and Machine Learning (ML).

AI is currently one of the most prominent topics in the IT industry. It has evolved from a buzzword to a central topic of discussion for technology and business leaders. AI is transforming how companies operate, compete, and ultimately sustain themselves, from streamlining processes to delivering highly personalized customer experiences.

AI's true power lies in its ability to analyze massive datasets and uncover insights that would be impossible for humans to generate within a reasonable timeframe. In today's data-driven world, this capability is invaluable. Whether it's predicting consumer behavior, detecting fraudulent transactions, or optimizing supply chains, AI is revolutionizing every aspect of business operations. AI's impact goes beyond enhancing operational efficiency. It's reshaping business models, creating new revenue streams, and redefining the competitive landscape.

As with any emerging technology, many organizations are eager to invest in AI to stay competitive. Companies that embrace AI are positioning themselves to lead the future, while those that hesitate to face the risk of disruption. However, adopting AI requires a far more substantial commitment and investment than the latest consumer products, like viral gadgets, like the latest Apple or Samsung products. The process involves carefully evaluating organizational needs, identifying relevant use cases, and developing a thorough implementation strategy that aligns with long-term business goals.

Machine Learning (ML) is a subset of AI focused on developing algorithms that enable computers to learn from data and make predictions. Unlike traditional AI, ML systems improve over time by analyzing patterns in data, such as identifying purchasing trends or forecasting sales.

AI and ML are transforming business operations by analyzing data, predicting trends, and automating tasks. While often used interchangeably, AI and ML have distinct roles. Artificial Intelligence (AI) encompasses technologies designed to mimic human intelligence, such as problem-solving and decision-making. Examples include expert systems and AI-powered chatbots that provide 24/7 customer support and personalized interactions based on user data.

In practice, AI and ML often work together. For instance, an AI-driven customer service platform may use ML to refine its responses based on customer interactions, enhancing its efficiency over time.

I have a complex relationship with Amazon, but recently, I noticed their AI-driven chatbot, which manages customer service inquiries, tracks orders, and answers questions. Using natural language processing (NLP) and ML, it continuously improves its responses. Amazon also utilizes AI to analyze user behavior for personalized recommendations, boosting satisfaction and reducing reliance on human agents. These advancements streamline my experience on their site, making navigation quicker and more efficient.

While AI aims to replicate human intelligence, ML focuses on data-driven learning and improvement. Understanding their differences can help businesses better utilize these technologies for enhanced customer service, decision-making, and innovative solutions.

In Artificial Intelligence Basics (Artificial Intelligence Basics: A Non-Technical Introduction, 2019), Tom Taulli offers a clear and concise introduction to AI, designed for readers with little to no background in the subject. The book demystifies complex AI concepts, such as machine learning and neural networks, making them digestible for a general audience. Taulli also explores the broader impact of AI on society, addressing important topics like ethical considerations, the future of work, and the technological advancements shaping industries.

Successful AI integration requires more than just technology. Transformation leaders must identify key areas where AI can enhance operations—whether it's customer service, supply chain optimization, or product personalization. Moreover, continuous assessment of AI's ethical implications and ensuring that its use aligns with organizational values is paramount.

2. The Emergence of Quantum Computing

Quantum computing marks a significant advancement in computational power, allowing for the resolution of complex problems that classical computers struggle with. By harnessing the principles of quantum mechanics, quantum computers can process information at speeds exponentially faster than traditional systems.

In 2019, Google announced it had achieved quantum supremacy, demonstrating that its quantum computer completed a specific task more quickly than the most powerful classical supercomputers. This milestone highlights the potential of quantum computing, although it currently applies to a limited range of problems. Companies such as IBM, Microsoft, and D-Wave are also actively advancing quantum hardware and software.

For Example, In the pharmaceutical industry, quantum computing could revolutionize drug discovery by simulating molecular interactions with unprecedented precision. While traditional drug discovery methods can be time-consuming and costly, quantum computing can evaluate millions of potential drug compounds much more quickly. This not only accelerates the development of new treatments but also reduces costs and enhances patient outcomes.

Quantum computing is set to transform various fields by addressing challenges beyond the capabilities of classical computers, despite the significant hurdles that remain, ongoing research and development promise to unlock new potentials and applications in the future.

3. Blockchain Technology

Blockchain technology ensures transparency and security through decentralized ledgers that record transactions across multiple computers. This technology is disrupting sectors by offering solutions to issues related to data integrity, fraud prevention, and efficient transaction processing. The emphasis here will be on how it is transforming the financial industry.

Blockchain technology is revolutionizing the finance industry by enhancing transparency, security, and efficiency. It provides a decentralized and immutable ledger that makes transactions more transparent and secure by using cryptographic techniques. This not only helps in reducing fraud but also lowers transaction costs by eliminating intermediaries like clearinghouses and settlement banks. Blockchain enables faster transactions, particularly for cross-border transfers, and supports automated, self-executing contracts known as smart contracts, which reduce manual intervention and errors.

In addition, blockchain fosters the growth of decentralized finance (DeFi) by allowing the creation of financial services that operate without traditional intermediaries. This broadens access to financial services and drives innovation with new financial products and services. Blockchain also enhances identity verification and fraud prevention through secure digital identity systems and improves compliance with streamlined auditing and reporting processes.

Moreover, blockchain facilitates the tokenization of assets, converting physical and digital assets into tradeable digital tokens. This increases liquidity and expands investment opportunities by allowing fractional ownership.

For businesses looking to adopt blockchain, the focus should be on evaluating where secure, verifiable transactions can add value and how they can integrate blockchain solutions without disrupting current operations. Blockchain-as-a-Service (BaaS) providers like Microsoft and IBM offer platforms that allow companies to experiment and implement blockchain without heavy upfront investment in infrastructure.

Despite its advantages, blockchain faces challenges such as regulatory uncertainty, scalability issues, and integration with existing legacy systems. Nevertheless, ongoing advancements promise to continue transforming the financial industry by improving how services are delivered and experienced.

4. Extended Reality (XR)

Extended Reality (XR) includes Virtual Reality (VR), Augmented Reality (AR), and Mixed Reality (MR), each offering immersive experiences and innovative ways to engage with digital content. These technologies are revolutionizing fields such as training, marketing, and entertainment.

Personally, I appreciate the benefits of XR; my family and I enjoy using our VR headset for engaging activities during family game nights and for exercise as an alternative to going to the gym. In the business world, VR is also making a significant impact. For example, a global manufacturing company might use VR to train employees on complex machinery. Rather than relying on traditional classroom methods, employees can use VR headsets to simulate operating machinery in a safe virtual environment.

This hands-on training helps employees develop and perfect their skills without the risks of real-world operations, leading to more competent staff and lower training costs. These diverse applications demonstrate how XR technologies are influencing both personal lives and organizational practices, highlighting their role in the ongoing digital transformation.

As discussed in earlier chapters, organizations must carefully evaluate the potential impacts and integration challenges when adopting disruptive technologies. This involves assessing how well the technology integrates with existing systems, understanding the required infrastructure, and managing any disruptions to current operations.

Effective change management is vital for the successful implementation of new technologies. Organizations need to prepare for changes in workflows, provide training on new systems, and clearly communicate the benefits and goals of the transformation.

Navigating digital transformation effectively requires a thorough understanding of disruptive technologies. By examining AI, quantum computing, blockchain, and XR, along with their practical applications, we can see how these innovations are not only disrupting traditional business models but also creating new opportunities for growth and efficiency. Staying informed about these technological advancements is crucial for maintaining a competitive edge and achieving future success.

5. IoT: Optimizing Operational Efficiency

IoT has the potential to connect devices, collect real-time data, and streamline operations across industries. In manufacturing, companies like General Electric (GE) have adopted IoT sensors to monitor equipment performance, predict maintenance needs, and reduce downtime. In retail, IoT has enhanced inventory management, allowing businesses to track products throughout the supply chain with unprecedented accuracy.

Businesses looking to integrate IoT should begin by identifying processes that can benefit from real-time data insights. Implementing IoT can drastically reduce operational costs by optimizing resource use, enhancing productivity, and improving customer experiences. The key to successful IoT adoption lies in scalability—starting with small pilot projects and gradually expanding across the organization as value is realized.

What are the Strategic considerations for sustaining gains in the digital era?

Sustaining gains in the digital era requires a strategic approach that addresses both immediate and long-term objectives. Embracing continuous innovation is foundational; organizations must cultivate a culture that encourages ongoing updates and improvements to digital tools and processes. This proactive investment in research and development helps them stay ahead of technological advancements and market shifts, ensuring competitiveness and enabling them to seize new opportunities.

Equally important is prioritizing cybersecurity. As reliance on digital systems grows, implementing comprehensive security measures, conducting regular vulnerability assessments, and staying abreast of emerging threats is critical to safeguarding data and maintaining trust.

Data-driven decision-making also plays a crucial role. By investing in advanced analytics and developing robust data governance strategies, organizations can enhance decision-making processes, improve forecasting accuracy, and personalize customer experiences more effectively.

Fostering a culture of agility is essential for adapting to the rapid changes characteristic of the digital landscape. Encouraging flexibility and responsiveness within teams, combined with agile project management practices, allows organizations to navigate new challenges and seize opportunities effectively.

Investing in talent development is vital for sustaining digital progress. Ongoing training and professional development equip employees with the skills necessary to leverage new technologies and drive innovation.

Concurrently, enhancing customer engagement through digital channels strengthens relationships, as tools for feedback and personalized interactions build customer loyalty and contribute to long-term success.

Optimizing digital infrastructure is key to supporting sustained growth. Regular reviews and improvements to IT systems ensure they remain robust, scalable, and aligned with evolving business needs, with solutions like cloud computing offering flexibility and cost efficiency.

Effective performance measurement and monitoring through clear metrics are important for tracking progress and identifying areas for improvement. Evaluating key performance indicators enables organizations to make data-driven adjustments to their strategies.

Building strategic partnerships with technology providers, vendors, and industry leaders further enhances capabilities and market reach. Such collaborations can accelerate digital solution implementation and provide access to innovative advancements.

Finally, ensuring compliance with regulations and establishing strong governance frameworks are essential for maintaining the integrity of digital practices. Staying updated on legal and ethical standards supports long-term success and upholds the credibility of digital operations.

Integrating Emerging Technologies: A Strategic Guide

Integrating emerging technologies such as AI, IoT, and blockchain into an organization's business model is a complex and ambitious undertaking with the potential for profound impact. Although this process is intricate, it offers substantial benefits, including enhanced efficiency, innovation, and a stronger competitive edge. To successfully navigate this integration, organizations must engage in thorough strategic planning, execute detailed implementation, and maintain a commitment to continuous evaluation.

As highlighted in the previous chapter, strategic planning is foundational to effective integration. This phase involves a comprehensive analysis of the current business environment and technology landscape.

Organizations must evaluate their existing processes, pinpoint opportunities where AI, IoT, and blockchain can add value, and establish clear, actionable objectives. Creating a strategic roadmap is crucial; it outlines both immediate goals, such as pilot initiatives, and long-term plans, including full-scale implementation and expansion strategies. This roadmap acts as a guiding framework, aiding organizations in prioritizing their projects and efficiently allocating resources.

Following the framework and strategies discussed earlier is vital for achieving successful outcomes. For AI, this means selecting appropriate tools and platforms, developing and training models, and integrating them into current systems. IoT integration involves deploying the right devices and sensors, ensuring reliable connectivity, and managing the data they generate. Blockchain integration requires choosing an appropriate platform and consensus mechanism, developing smart contracts, and ensuring seamless functionality across all components. Each technology demands a customized approach to address specific challenges, necessitating a structured and methodical implementation process.

Ongoing evaluation is critical to ensure that the integration meets its goals and delivers the anticipated benefits. This involves setting performance metrics and key performance indicators (KPIs) to track progress and effectiveness. Collecting feedback from stakeholders and users helps identify areas for improvement and necessary adjustments. Regular reviews and modifications based on performance data are essential for optimizing processes and technologies, ensuring they continue to align with the organization's goals and adapt to the evolving business landscape.

Integrating AI, IoT, and blockchain into a business model is a strategic endeavor that requires careful planning, precise execution, and continuous assessment. By following the comprehensive roadmap outlined in this book, organizations can successfully navigate the complexities of these technologies, unlocking their full potential and driving significant business success. This approach not only helps in overcoming the challenges associated with integration but also unlocks the full potential of these technologies, driving significant business success and ensuring a sustainable competitive edge.

Chapter 10

Conclusion - Navigating the Future of Digital Transformation

As we conclude our deep dive into digital transformation, it is clear that navigating this journey demands a comprehensive and multifaceted approach. Digital transformation extends far beyond simple technological upgrades; it represents a fundamental shift in how organizations operate, engage with customers, and create value. Throughout this book, we have explored the complex nature of digital transformation, the forces driving it, and the obstacles organizations may face. This final chapter will consolidate the key insights from each section and offer a forward-looking perspective on sustaining and advancing in the digital age.

Digital transformation is a profound change that involves not just the adoption of new technologies but also a strategic overhaul of organizational culture, operations, and business models. At its heart, it is about leveraging digital technologies to fundamentally alter how organizations function, interact with customers, and deliver value. The primary drivers behind this transformation include shifting customer expectations, rapid technological advancements, and intense competitive pressures, all of which underscore the need for organizations to continuously adapt and innovate.

The challenges associated with digital transformation, as detailed in this book, are significant and multifaceted. Organizations often encounter a range of obstacles, from technological constraints and resource limitations to internal resistance and external market forces. Identifying these challenges early and addressing them proactively is crucial to ensuring a successful transformation process.

In exploring the technologies and strategies that underpin digital transformation, we have examined a wide array of tools and approaches. Technologies such as AI, IoT, and blockchain offer unique capabilities that can drive substantial change. Evaluating these technologies requires a thoughtful assessment of their potential impact, alignment with organizational goals, and the ability to deliver a meaningful return on investment. Effective prioritization of initiatives based on feasibility and ROI is essential for navigating the complex landscape of digital transformation.

Successful execution of digital transformation initiatives hinges on effective leadership and change management. As discussed, leaders play a pivotal role in guiding transformation by articulating a clear vision, cultivating a culture of innovation, and managing the complexities of change. Strong leadership is crucial for overcoming resistance and aligning the organization with its strategic objectives.

Data-driven decision-making is another critical component of successful transformation. The ability to leverage data analytics to inform strategies and decisions provides a powerful advantage. Coupled with robust data governance and quality management practices, data analytics enhances decision-making processes and ensures that strategies are based on accurate and actionable insights.

Collaboration and continuous improvement are integral to the success of digital transformation. By forming strategic partnerships and building ecosystems, organizations can access external expertise and resources, which drives innovation and expands capabilities. Effective management of technology vendors and the fostering of collaborative relationships are vital for creating a cohesive and adaptable digital ecosystem.

Measuring success and refining strategies based on performance metrics are essential for achieving long-term success. Establishing clear KPIs and metrics allows organizations to track progress, assess the impact of transformation efforts, and identify areas for improvement. Embracing a continuous improvement mindset ensures that organizations remain flexible and can sustain their digital transformation benefits over time.

Finally, we have explored emerging trends and their implications for digital transformation. Staying informed about technological advancements and market changes is crucial for maintaining a competitive edge and driving future growth. Strategic planning for sustaining digital transformation gains involves not only leveraging current technologies but also anticipating and preparing for future developments.

In summary, digital transformation is an ongoing journey, not a final destination. While the process is intricate and challenging, the potential for significant impact and long-term benefits makes it a worthwhile endeavor. Organizations that approach this journey with a strategic mindset, a commitment to continuous learning, and a readiness to adapt will be well-positioned to thrive in the digital era.

By applying the insights and strategies outlined in this book, organizations can effectively navigate the complexities of digital transformation, unlock new opportunities, and achieve sustained success in a constantly evolving business landscape.

References

Artificial Intelligence Basics: A Non-Technical Introduction. (2019). Apress.

BBC. (2018, March 21). *Cambridge Analytica: Facebook row firm boss suspended.* Retrieved from BBC: https://www.bbc.com/news/uk-43480048

BBC. (2020, October 16). *British Airways fined £20m over data breach.* Retrieved from BBC: https://www.bbc.com/news/technology-54568784

Blackberry. (2024, October 23). Retrieved from Wikipedia: https://en.wikipedia.org/wiki/BlackBerry#:~:text=BlackBerry%20is%20a%20discontinued%20brand,was%20licensed%20to%20various%20companies

e-careers. (2024, April 18). *10 businesses that failed to adapt.* Retrieved from e-careers: https://www.e-careers.com/connected/10-businesses-that-failed-to-adapt

Jeffrey M. Hiatt, T. J. (2003). *Change Management.* Prosci.

Kissflow Inc. (2024). *Digital Transformation.* Retrieved from kissflow: https://kissflow.com/digital-transformation

Kitty Paul, T. A. (2018). *Kitty Paul, Tom Arnold, Marwa Rashad and Stephen Kalen.*

KPMG LLP. (2024, May). *2023 KPMG US Technology Survey Report.* Retrieved from kpmg: https://kpmg.com/us/en/articles/2023/us-tech-survey-2023.html

McKinsey & Company. (2018, October 29). *Unlocking success in digital.* Retrieved from Mckinsey.com: https://www.mckinsey.com/search?q=unlocking%20success

Nanda, A. (2017, November 16). *Digitization of Walmart*. Retrieved from Digital Initiative: https://d3.harvard.edu/platform-rctom/submission/digitization-of-walmart/

Nestlé reports full-year results for 2020. (2021). Retrieved from Nestle: https://www.nestle.com/sites/default/files/2021-03/creating-shared-value-report-2020-en.pdf

Rob Llewellyn. (2024, June). *The 50 Best Examples of Business Model Transformation*. Retrieved from Rob Llewellyn: https://robllewellyn.com/business-model-transformation-2/#:~:text=Amazon%20transformed%20its%20business%20model,revenue%20through%20its%20platform%20services

TQM's Challenge to Management Theory and Practice. (1994, Jan). Retrieved from MITSLOAN: https://sloanreview.mit.edu/article/tqms-challenge-to-management-theory-and-practice/

Uber Hid 2016 Breach. (2017). Retrieved from nytimes: https://www.nytimes.com/2017/11/21/technology/uber-hack.html

Why The Commanders' Owner Thinks $6 Billion For An NFL Team Is A Bargain. (2024, October). Retrieved from Forbes: https://www.forbes.com/sites/bernardmarr/2018/08/09/ibm-and-maersk-launch-blockchain-based-platform-for-global-trade/

About The Author

Elizabeth is a distinguished digital transformation leader and a highly sought-after speaker, currently serving as the lead consultant at Privy Consulting, a boutique IT consultancy and training firm. With over a decade of experience, she has been instrumental in guiding organizations across a wide range of sectors—including financial services, information technology, education, and healthcare—in the successful implementation of disruptive transformation programs. Her work is renowned for driving substantial growth, enhancing business efficiency, and fortifying corporate resiliency.

Elizabeth holds a Master of Business Management (MBA) with a focus on General Management, complemented by a suite of prestigious certifications that highlight her commitment to excellence. Her accolades include advanced training from Harvard Business School in Distributive Strategy, providing her with a profound understanding of market disruption theory and strategic innovation. Additionally, she is a Certified Business Analysis Professional (CBAP) recognized by the International Institute of Business Analysis, a Certified ScrumMaster (CSM), and holds certifications in Strategic Management and Leadership.

Beyond her theoretical expertise, Elizabeth is actively engaged in merit-based professional societies, such as the Chartered Management Institute (CMI), further showcasing her dedication to continuous professional growth. Her extensive experience and unwavering commitment to advancing the field of digital transformation solidifies her status as a leading thought leader and a transformative force in the ever-evolving digital landscape.